THE DANGERS
OF INTELLIGENCE
AND OTHER
SCIENCE ESSAYS

THE DANGERS
OF INTELLIGENCE
AND OTHER
SCIENCE ESSAYS

Isaac Asimov

BOSTON

Houghton Mifflin Company

1986

Library of Congress Cataloging-in-Publication Data

Asimov, Isaac, date.
The dangers of intelligence and other science essays.

1. Science—Popular works.   I. Title.
Q162.A77   1986   500   86-249
ISBN 0-395-41554-3

Printed in the United States of America

S 10 9 8 7 6 5 4 3 2 1

Dedicated to the memory of
Theodore Sturgeon (1918–1985)
and Larry Shaw (1924–1985)

CONTENTS

INTRODUCTION

IN 1981, MY OLD FRIENDS at Houghton Mifflin (with whom I've been publishing books for over a quarter of a century) published my book *Change!*, which consisted of seventy-one short essays, each representing an imaginative look at some aspect of a possible future. (All futures are no better than "possible," for we don't really know what tomorrow holds.) The essays were taken from *American Way*, the in-flight magazine of American Airlines, which has printed my stories in every issue now for over eleven years.

Even while *Change!* was in press, and after it was published, my essays continued to appear in every issue of *American Way*, for its editorial staff was sufficiently good-natured and easygoing to let that happen. Indeed, at the start of 1985, the magazine went from monthly to biweekly publication, and they simply told me to rev up my word processor. Now I do twenty-six essays a year instead of twelve. It should come as no surprise to my long-suffering Gentle Readers, then, to learn that I have accumulated another group of essays and have distributed them lovingly in the pages of a book. This time, however, there are seventy-two essays rather than seventy-one, just to make matters a little different.

The essays are not printed in the order of their appearance in *American Way*, because that would make the book too chaotic. You see, I have a free hand to write about any subject I like, and it keeps my interest piqued to the maximum if, in my biweekly essays, I move about wildly from subject to subject, following no particular plan other than that of yielding to impulse. However, having these essays appear one at a time in a magazine containing many other items as well is one thing; having them all appear one after the other in a single book is another. In order to impose some sort of sense upon them, I have tried to put them in a semblance of order, with one essay leading to the next in a reasonable way whenever that was at all possible.

As I warned you in *Change!*, some of these futures may contradict others, or overlap others for that matter. What's more, the essays in this book were written over a period of five years, and in a few cases time has caught up with them in one way or another. Whenever a projected occurrence has become less likely or has been eliminated or involves anything new of interest that I know of, I have told you about it in a note following the essay.

And if, when you finish reading this book, you feel you haven't had enough, just hang on. Assuming I continue in good health, and the staff at *American Way* retains its easygoing, good-natured spirit, and Houghton Mifflin Company remains willing as well—then, in three years or so, I may have a third volume in this series for you.

ISAAC ASIMOV
August 1985

THE DANGERS
OF INTELLIGENCE
AND OTHER
SCIENCE ESSAYS

1 THE TWENTY-SIX LETTERS

THE RATE OF ILLITERACY in the world is decreasing, the percentage of adults who cannot read is going down—but only slowly. It is not going down as fast as the world population is going up. In 1970, for instance, it was estimated that there were 760 million adult human beings in the world who could not write and who could not read what others wrote. That was about 21 percent of the total population of the world. By 1980, only 18.5 percent of the total population of the world was illiterate, but the population had gone up so much that the actual number of illiterates was 814 million. In ten years, in other words, the number of illiterates had increased by 54 million, a number equal to the population of France.

Does it matter? It certainly does. Ever since writing was invented, about five thousand years ago, it has been the essential method for recording and transmitting information, and to a large extent people who cannot read or write are information-blind. To be sure, there is always speech itself, but the spoken word is far more limited than the written word. Until recently, it was impossible to store speech as such, and it has always been very difficult to recall the spoken word accurately. Nowadays, when we *can* store the written word in recordings of one form or another, it remains less convenient to retrieve it and to work with it in the form of sounds alone. Inevitably, in order to make use of sounds, however cleverly recorded, we convert them into writing.

There have been improvements in the technology of writing (the printing press, the typewriter, the word processor) but these inventions have only improved the manner in which speech is

1

coded into marks on a suitable surface. We now have quicker and easier and more efficient ways of writing, but what is finally produced remains writing. And someone who can't read the written code as formed by hand can't read it when it is printed, typewritten, or placed on a television screen either.

In addition to the total illiterates there is an equal number who while not technically illiterate, can read and write only very slowly and can handle only the simplest words. They are "functionally illiterate" in that they cannot write with sufficient ease to work in a technological society. They can perform only what is called unskilled labor, and even then, they must be instructed with particular care, since one can't depend on them to pick up guidance from labels, street signs, store posters, or any of the many other casual information-bearing items that saturate our world and that we assume everyone can make use of.

At least one out of three adults in the world can't function usefully in a technological society. This badly limits the developing nations particularly, where the illiteracy rate is highest. It makes it that much more difficult for them to take advantage of modern technology in order to build a better life for themselves. Nor are the industrial nations entirely free of this limitation. While the total illiteracy rate in the United States is very low, there are millions of adult Americans who are functionally illiterate just the same, and who as a result are condemned to the lowest rung on the economic ladder. Their inabilities also exert a downward drag on the economy as a whole. What can we do about it?

About 1400 B.C., the Phoenicians invented the first alphabet and writing was enormously simplified. All other alphabets, however strange they may seem, are thought to be modifications of this great original, and of all the modifications, the Roman alphabet—the one in which this essay appears—is the most widespread and the most used. Might it not help the world generally if all of it used an alphabet? (China, notably, uses ideographs instead.) Might it not in fact be helpful if the whole world used the same alphabet—the Roman? (The Soviet Union, notably, uses the Cyrillic alphabet instead.) To be sure, the Roman alphabet is not perfectly adapted to every language, not even to English. Thus, the letters c, q, and x are superfluous, since c does nothing that k and s can't do, and q could be replaced by k, and x by ks or z. And meanwhile there are no single letters that represent the common English consonantal sounds of *sh, ch,* and *th.*

How useful it might be, then, were a world committee of scholars to get together and devise a world alphabet. It would not necessarily be one that would suit every language perfectly, since that might require far more than the traditional twenty-six letters, but it might at least make it possible to record every language with *reasonable* fidelity. In that case everyone in the world, when learning to read and write, could make a beginning with the same alphabet. Once a person had learned to read and write in any one language, he would also be able to pronounce words in any other language at sight, even if he did not understand them.

2 DIAL VERSUS DIGITAL

THERE SEEMS NO QUESTION but that the clock dial, which has existed in its present form since the seventeenth century and in earlier forms since ancient times, is on its way out. More and more common are the digital clocks that mark off the hours, minutes, and seconds in ever changing numbers. This certainly appears to be an advance in technology. You will no longer have to interpret the meaning of "the big hand on the eleven and the

3

little hand on the five." Your digital clock will tell you at once that it is 4:55. And yet there will be a loss in the conversion of dial to digital, and no one seems to be worrying about it.

When something turns, it can turn in just one of two ways, clockwise or counterclockwise, and we all know which is which. Clockwise is the normal turning direction of the hands of a clock and counterclockwise is the opposite of that. Since we all stare at clocks (dial clocks, that is), we have no trouble following directions or descriptions that include those words. But if dial clocks disappear, so will the meaning of those words for anyone who has never stared at anything but digitals. There are no *good* substitutes for clockwise and counterclockwise. The nearest you can come is by a consideration of your hands. If you clench your fists with your thumbs pointing at your chest and then look at your fingers, you will see that the fingers of your right hand curve counterclockwise from knuckles to tips while the fingers of your left hand curve clockwise. You could then talk about a "right-hand twist" and a "left-hand twist," but people don't stare at their hands the way they stare at a clock, and this will never be an adequate replacement.

Nor is this a minor matter. Astronomers define the north pole and south pole of any rotating body in such terms. If you are hovering above a pole of rotation and the body is rotating counterclockwise, it is the north pole; if the body is rotating clockwise, it is the south pole. Astronomers also speak of "direct motion" and "retrograde motion," by which they mean counterclockwise and clockwise, respectively.

Here is another example. Suppose you are looking through a microscope at some object on a slide or through a telescope at some view in the sky. In either case, you might wish to point out something to a colleague and ask him or her to look at it, too. "Notice that object at eleven o'clock," you might say—or five o'clock or two o'clock. Everyone knows exactly where two, five, or eleven—or any number from one to twelve—is located on the clock dial, and can immediately look exactly where he is told.

(In combat, pilots may call attention to the approach of an enemy plane or the location of antiaircraft bursts or the target, for that matter, in the same way.)

Once the dial is gone, location by "o'clock" will also be gone, and we have nothing to take its place. Of course, you can use directions instead: "northeast," "southwest by south," and so on. However, you will have to know which direction is north to begin with. Or, if you are arbitrary and decide to let north be straight ahead or straight up, regardless of its real location, it still remains true that very few people are as familiar with a compass as with a clock face.

Here's still another thing. Children learn to count and once they learn the first few numbers, they quickly get the whole idea. You go from 0 to 9, and 0 to 9, over and over again. You go from 0 to 9, then from 10 to 19, then from 20 to 29, and so on till you reach 90 to 99, and then you pass on to 100. It is a very systematic thing and once you learn it, you never forget it. Time is different! The early Sumerians couldn't handle fractions very well, so they chose 60 as their base because it can be divided evenly in a number of ways. Ever since, we have continued to use the number 60 in certain applications, the chief one being the measurement of time. Thus, there are 60 minutes in an hour.

If you are using a dial, this doesn't matter. You simply note the position of the hands and they automatically become a measure of time: "half past five," "a quarter past three," "a quarter to ten," and so on. You see time as space and not as numbers. In a digital clock, however, time is measured *only* as numbers, so you go from 1:01 to 1:59 and then move directly to 2:00. It introduces an irregularity into the number system that is going to insert a stumbling block, and an unnecessary one, into education. Just think: 5.50 is halfway between 5 and 6 if we are measuring length or weight or money or anything but time. In time, 5:50 is nearly 6, and it is 5:30 that is halfway between 5 and 6.

What shall we do about all this? I can think of nothing. There is an odd conservatism among people that will make them fight to the death against making time decimal and having a hundred minutes to the hour. And even if we do convert to decimal time, what will we do about "clockwise," "counterclockwise," and locating things at "eleven o'clock"? It will be a pretty problem for our descendants.

---

3 THE DANGERS OF INTELLIGENCE

IT IS AN INDISPUTABLE FACT that human beings dominate the Earth. They can drive plants and animals to extinction without any difficulty, and they can change the face of the planet and alter the environment to suit their own needs and whims. No species of plant or animal that has ever lived could even begin to do these things in the way we can. And it is equally clear that these abilities are the result of our intelligence. No other species of living thing on this planet has (or has had in ages gone by) our kind of intelligence.

I suppose that few people are willing to give up the general intelligence that human beings have, or are dissatisfied with our ability to establish increased comfort and security for ourselves. But is intelligence really a good thing *in the long run?* Does it really help living things to survive?

You might think it does. You might suppose it is pretty obvious that living species with better brains can outthink and outsmart other species and therefore outlive them—but is it? Next to ourselves, chimpanzees and gorillas are the brainiest of the land animals, but they are not notably successful in the battle for survival. To be sure, their present problem lies in their competition for living space with human beings, and they can't win there, but they weren't terribly successful even before people began to press directly upon them. Neither were elephants, which are also pretty brainy (their brains are larger than ours, in fact).

If you want to compare chimpanzees and elephants with a truly successful species, consider the common rat. It lives right in human space, competes directly with us, withstands all the most ferocious human attempts to wipe it out, and flourishes. That's *successful*. Rats are pretty smart for their size, but they are not as smart as chimpanzees. What gives rats survival ability is not so much their intelligence as their size (they are small and hard to find), their dietary habits (they eat a wide variety of food and are hard to starve out), and, most of all, their fecundity (they have many young and can quickly replace any losses).

And if you think that rats are successful, consider roaches, which also live and compete with human beings, withstand all human attempts to destroy them, and flourish—yet have virtually no brains at all. But then, compared with rats, roaches are still smaller, have an even broader diet, and are yet more fecund. In fact, I wonder whether the ancestors of *Homo sapiens* paved the way for human conquest of the world through the development of larger brains or a wider diet. Our food habits, rather than our intelligence, may have been the initial secret of our success. Of course, once intelligence passes a certain point and enables a species to develop an elaborate technology, it begins to count for a great deal.

It is technology—the use of fire, elaborate tools, and new and subtle inventions—that makes all the difference, not intelligence alone. (Dolphins may be highly intelligent, but so what? They have no technology.) But then, in our own case, we find that we have reached the point where technology—by increasing the food supply, security against predators, medical knowledge, and so on—has made it possible for us to multiply in an unrestrained fashion, to consume Earth's resources faster than

7

they can be replaced, and to poison its soil, water, and air faster than they can be cleaned. We can even physically destroy our planet with nuclear warfare. Is this process inevitable? Is intelligence useless for survival till it passes a certain point, and then must it bring world domination, followed by suicide?

People say, "If there are other intelligences out there, why haven't they gotten in touch with us?" Can it be that the other intelligences out there either haven't evolved enough to build a technology with which to reach us—or have, and have then destroyed themselves *before* they could reach us? Thoughts like these could cause us to despair, to see the suicide of humanity as absolutely unavoidable. And if we do despair and don't fight desperately to prevent the destruction, it *will* be inevitable.

That is why I would like to see a strong attempt made to detect signals from space that might indicate the presence of another working civilization. Surely any civilization that can send out signals we can receive is at least as advanced as we are. Very likely, it has far surpassed us. Such signals would at once give us an all-important message, even if we understood absolutely nothing of what was being said. The message would be, "We have a technology more advanced than yours and we have managed to survive. Take heart! You can do so, too." It is encouragement we badly need.

4 LITTLE THINGS

ABOUT 100 B.C., in India, someone thought of hanging a loop of leather from each side of a saddle so that a big toe could be placed in each loop. This was the first toe stirrup, and it helped steady a rider on a horse. Considering that horses had been ridden for nearly 2,000 years, you'd think someone would have thought of it sooner.

The idea spread to China, where the climate was colder and people did not ride horses barefoot. About A.D. 400, someone thought of expanding the toe stirrup into a metal loop large enough to hold the shoe. This device became the foot stirrup. A little thing, but what a difference it made. Until then, a mounted warrior could thrust a spear only from a stationary horse. If the warrior made his thrust when the horse was running, it would run right out from under him and the rider would be thrown. With the aid of the stirrup, the rider could hold himself firmly in the saddle and the entire weight of the running horse would be behind the spear thrust. No foot soldier could stand against it.

The idea of the stirrup spread rapidly to the nomads of Central Asia. They put an end to the Roman legions who had ruled the Mediterranean world for over five centuries, and the Huns rode irresistibly across Europe. About 730, the Frankish leader Charles Martel raised a whole army of cavalry on large horses originally bred in Persia. Both rider and horse wore armor and there were stirrups, too. On this "heavy cavalry," Martel's grandson Charlemagne built his empire. But it cost money to have a horse and armor and to undergo the training. Only the landed aristocracy could do it, and only they had power as a result. And so feudalism was fastened on western Europe for six centuries, until new weapons—the pike, the crossbow, and the cannon—were invented.

Here's another little thing. The Mediterranean world had led in civilization and power for thousands of years. They had developed agriculture, for instance. Their agricultural methods didn't work in the heavy, damp soil of northern Europe, however, and the north therefore remained forested, underpopulated, and weak. Eventually, a new kind of plough, with wheels, was invented that could cut into the heavy, sticky soil, but it took a great deal of force to pull the plough through that soil. Donkeys were too small and weak, oxen too slow. Horses were the only animals strong enough to do the job, but harnessing them to the plough in the same way that oxen were harnessed didn't work. Horses were different from oxen anatomically, and the harness pulled against their windpipe. The harder the horse pulled the plough, the less he could breathe. It meant a horse could only pull with about one-fifth of his maximum force.

But somewhere in the east, a horse collar was invented. This was a padded ring of tough material that was placed at the base of a horse's neck and rested against its shoulders. The harness was attached to the horse collar, which meant that the horse pulled with its shoulders and not with its windpipe. This invention reached northern Europe in about the year 1000, and about that time, too, the notion of horseshoes came in to protect the horse's sensitive hooves from rocks and other obstructions. Horses could be used to pull the new plough; northern farming became efficient; the food supply shot up; the population (especially among the well-fed aristocracy) multiplied. There were too many aristocrats fighting over a fixed amount of land, so France and Germany grew thick with endless battling. The Crusades began with the notion of regaining the Holy Land, yes, but also with the notion of getting rid of some of the quarrelsome nobles and letting them exert themselves against the infidel. From the East, the Crusaders brought back civilization and the stage was set for what would eventually be a three-century domination of the whole world by the northern Europeans. All from a stirrup. All from a horse collar.

You can't always predict what will come out of little things or know what's happening while it's happening. In 1948, the transistor was developed. That started a rapid and massive development of solid-state physics. From the transistor came the microchip, which made possible the home computer and the industrial robot, and where that's going to take us, we still can't see clearly. All from a transistor. Or a special valve is invented that makes it possible to build spray cans out of thin, cheap aluminum with no significant danger of explosion. Suddenly spray cans proliferate, and just as suddenly there arises the possibility that the fluorocarbons used to force the spray outward may accumulate in the upper atmosphere and put an end to the ozone layer that alone protects life against killing radiation from the Sun. All from a little valve.

Where's the next little thing coming from? Will we recognize it as important when it comes? Will we be prepared for the great changes that may follow? And the possible great dangers?

NOTE: *I might have mentioned nonhuman "little things." There was the germ of a new virulent strain of plague that swept the Earth with the Black Death in the 1300s and wiped out perhaps a third of humanity. And (possibly) there is the virus that is spreading* acquired immune deficiency syndrome (AIDS) *now.*

---

5 THE GREEN ENEMY

HUMANITY HAS THREE CLASSES of living nonhuman enemies. First, there are the great predators: lions, bears, sharks, and so on. We treasure stories of Samson rending a lion and we shudder over the movie *Jaws.* Actually, however, those poor animals have been outclassed for thousands of years and could be driven to extinction with very little trouble if humanity really put its mind to it. Second, there are the invisible parasites: the viruses, bacteria, protozoa, worms, and so on, that in one way or another live at our expense and interfere with our health. These are far more dangerous than the large predators, and we need only compare the Black Death of the fourteenth century with anything man-eating tigers could do. In the last century and a quarter, however, we have learned ways of dealing with these disease producers, and the danger has vastly diminished. That leaves the third group: unwanted plants, or weeds. With very few exceptions, these are not apparently dangerous in themselves and are certainly not dramatic, for they do nothing but grow. And yet in some ways they are the most insidious and dangerous enemy of all.

What distinguishes a weed from other plants? Only that factor of uselessness. If you were growing dandelions and wanted them desperately, you would consider roses as weeds. Some weeds are merely despised as unsightly. They are not pretty or orderly plants, and they make an area look rotten. Others produce pollen that increases the sufferings of those with hay fever or are even actively poisonous if eaten. For the most part, though, weeds are dangerous merely because they grow along with crops, snaring sunshine and absorbing water and mineral nutrients that human beings feel should go to those crops. Left to themselves, such weeds are bound to lower crop production drastically.

It is for this reason that an essential part of farming is the act of weeding, the careful, painstaking removal of unwanted plant competition, one plant at a time. It is backbreaking, never ending, and wholly undramatic work, but it is absolutely essential, for without it, humanity would be struck with a devastating famine. In reverse, if we could find a more efficient way of removing weeds than digging them up one by one, the food resources of the world could be greatly increased. The contemporary tactic is to make use of chemicals of one sort or another. The great advance in this direction came with the introduction of a weed killer (or herbicide) named 2,4-D in 1945. Since then, dozens of different herbicides have been developed, and hundreds of thousands of tons of them are produced each year.

A herbicide must stop plants without harming animals. All green plants depend on photosynthesis, the manufacture of plant tissue through use of the energy of sunlight; and the complex system for carrying this out is located within the tiny "chloroplasts" present within the cells of leaves and other green tissues. It would be useful, then, to have chemicals that would interfere with some key step or steps in photosynthesis. This would cause the plant to starve to death but would have no direct effect on animals, who don't use the process. And, as a matter of fact, most herbicides in use work in this fashion. Second, you must find a chemical that will interfere with photosynthesis in weeds, but *not* in crops. Surprisingly enough, this is possible. Sometimes leaves are drastically different in design. Grass leaves are thin and vertical, whereas the leaves of weeds in lawns are usually broad and horizontal. This gives the weeds an advantage in collecting sunlight—and a disadvantage in collecting herbicide spray. Sometimes crops are tall and weeds are short, or crops start early in the season and weeds come later (or vice versa), and the spray can be applied at a particular height, or at a particular time, so that the weeds are affected and the crops are not.

What is most needed, however, is the kind of detailed information about the photosynthetic mechanism that we don't as yet have. The mechanism is so complicated, after all, that it is bound to vary in many ways from one species of plant to another. What if we could design a specific herbicide, one that was specially adapted for penetrating the cell of one species but not another? Or, once in the cell, a herbicide designed to interfere with a particular enzyme that was present in one species but not in another or was particularly susceptible in one species but not in another? We might then have an entire armory of herbicides, each with its particular job. Since each would be specifically designed for a specific task, each could be used in minimal quantities. This would reduce the kind of undesirable side effects that take place when herbicides are used in large quantities in order that they might exert their killing effect by sheer overwhelming force, since they are not particularly well adapted to their job.

Of course, there will always be some natural resistance, and it is the resistant plants that will survive and multiply, so the armory of herbicides will have to be continually brought up to date—but that will be a small annoyance compared to the benefits of a large increase in crop production and a decrease in undesirable side effects.

NOTE: *I'm not really keen on killing living things, and who are we to decide certain plants are "weeds"—but as long as human beings insist on multiplying their numbers to the point of mass starvation, we have no choice but to try to increase our food supply. See the next essay for the other side of the coin.*

6 THE DYING FORESTS

UNTIL 1800, THE GREAT FUEL of the world was wood. Coal was used, but only to a trifling extent. Fats and oils from animals and plants were used only for candles and lamps, while petroleum and natural gas were used only where some leaked out of the ground. Since then, of course, the industrial world has come to use coal by the billions of tons, oil by the billions of barrels, natural gas by the billions of cubic feet. Wood, it might seem, is not used at all. I myself have never seen wood burning except occasionally in ornamental fireplaces—for show rather than for use.

The disappearance of wood, however, is an illusion born of living in a large city in a developed nation. In the countryside of even developed nations wood is still used, and in undeveloped nations wood is still, even as the twentieth century dwindles to an end, the *chief* fuel. About 2 billion people in the poorer nations of the world must depend on wood almost entirely for their daily energy needs. To be sure, the people of the poor nations don't use much energy per capita by American standards, but their total requirement, thanks to their numbers, is not trifling. There is only one source from which they can get wood, moreover, and that is trees. Trees must be chopped down for the purpose. As population increases (and it is increasing rapidly in all the poor nations), more energy is needed, and still more trees must be chopped down. Of course, trees are a renewable resource, for new trees keep growing spontaneously. They are not totally renewable, however, since trees can be and, these days, *are* being cut down faster than new ones can grow. Nor is the need for fuel the only driving force behind the destruction. With increasing population, more food is needed, and this usually means more land must be found on which to grow crops. Such new land is most frequently made available by chopping down sections of forests.

The forests being destroyed for fuel and land are located mostly in the tropical countries, where they are disappearing with frightening quickness. The estimate is that the tropical forests are being cleared at the rate of sixty-four acres *per minute*. What is taking place is an appalling catastrophe. Fuel is running short as a result. Already in 1981, it was estimated that as many as 96 million people could not get enough fuel wood to meet their minimum needs for cooking and heating. By the end of the century, 2,400 million people may be in that fix. Nor can we imagine that wood can be replaced by more sophisticated energy sources. Developing such sources would take time and capital investment, which the poor nations don't have, and charitable donations of coal and oil from an industrial world (that feels itself to be increasingly in danger of shortages, too) are not in the cards. For that matter, the industrial nations are killing their own forests with the acid rain that results from the burning of too-impure coal and oil.

As it happens, trees perform many functions other than the simple manufacture of wood. Their roots hold soil in place more efficiently than the roots of other plants do, and they absorb water, preventing a too-rapid run-off. What is more, the copious rains (which give the rain forests of the tropics their name) leach the soil and leave it poor in minerals. The tropical forests are adapted to this situation and grow well under these conditions. Other plants would not. What's more, forests discharge water into the atmosphere through their leaves in great quantity. This water is vaporized, and the process absorbs much heat that would otherwise serve to warm the ground.

Cut down the forests, and the farming will work well for a few seasons and then dwindle as the fertility of the soil is worked out by farmers who cannot afford fertilizers. The drenching rains will gully the fields and wash away the soil, which will grow hotter and drier until badlands are formed. On the whole, deserts will expand along their every rim (the process is called "desertification"), so that fertile land will steadily shrink in area and the food supply will dwindle. If the population continues to grow, you can easily imagine the inevitable result.

Then, too, forests harbor millions of species of plants and animals that may be wiped out with them. Most of those species have not yet been studied, but the number of useful products in the way of medicinals, for instance, that we have already obtained from otherwise obscure life forms makes it reasonable to suppose that enormous numbers of beneficial discoveries remain to be made in this respect. If the myriad forest-nurtured life forms die, we stand to lose many benefits that, in most cases, we don't even know we have.

The dwindling of the forests will change our climate for the worse, our food supply for the worse, the diversity and benefits we might gain from other life forms for the worse. It will change nothing at all for the better. Maybe, as Joyce Kilmer says, "only God can make a tree," but if so, human beings had better start helping. We've got to plant more trees than we cut down.

7 THE 8,000-YEAR CALENDAR

ABOUT SEVENTY YEARS AGO, an American astronomer, Andrew Ellicott Douglass, who worked in Arizona, began to study wood. Old pieces of wood were perfectly preserved in Arizona's dry climate, and what he studied were the tree rings.

Every summer, wood grows rapidly if the weather is suitable over the year generally; slowly, if it is not. This pattern of rapid and slow growth produces the effect of rings, one ring for each year. If a summer is unusually cool or unusually dry, the growth ring is narrow. A warm, wet summer, on the other hand, would produce a wide growth ring. Douglass was at first interested in seeing whether there was a regular alternation of fast growth and slow growth with the sunspot cycle, but there wasn't, and he found himself growing interested, instead, in climates of the past as recorded in the tree rings.

In a living tree, he would find a particular pattern of rings, wide and narrow, that might extend back a hundred years. (It is not necessary to kill a tree to do this. A core of wood can be bored from bark to center, taken out, and studied. The tree will heal.) Suppose you studied a piece of wood that you suspected was part of a tree cut down a few decades ago. Its ring pattern would fit an older portion of the pattern of the living tree, and counting back to the place where the pattern began to fit, you might find that the wood came from a tree that was cut down perhaps as many as 34 years ago. This tree might have been growing before the living tree was, and you could follow the pattern back 162 years, perhaps, from the present.

A still older piece of wood could be matched against the 162-year-old pattern, and the pattern could then be pushed back to still earlier times. By 1920, Douglass had worked out a pattern that stretched back to about A.D. 1300. It was the first method worked out that could date early events accurately in the absence of historical records. If traces of an ancient Indian village were discovered and wood had been used in constructing a house, then the date when that house had been built could be worked out quite accurately from the ring pattern. What's more, determining the date didn't depend entirely on matching different pieces of wood. It is possible to find a single piece of wood that gives one a calendar thousands of years long.

In the West, there grow bristlecone pines that are the oldest known living things. One of them, growing in eastern Nevada, was cored and found to be 4,900 years old. It was a seedling in 3000 B.C. and was several centuries old before the first pyramid was built in Egypt. It can be used as a one-piece, five-thousand-year-old calendar to check the tree-ring pattern. In fact, the work of Edmund Schulman with these trees in the 1950s carried the calendar back nearly eight thousand years. Lately, tree-ring data have begun to be used in another way — to study volcanic eruptions. The eruption of Mount Saint Helens a few years ago and the later eruption of a volcano in southern Mexico have perhaps made Americans a little more volcano-conscious than before, so that questions about them have arisen.

Volcanoes spew dust into the atmosphere, and if the volcanic eruption is violent enough, cubic miles of dust can be lofted high into the stratosphere. This dust can spread out in a world-circling layer and sometimes not settle back to the ground for several years. The dust in the upper atmosphere reflects sunlight, and less sunlight reaches the Earth's surface than would do so if the dust weren't there. For that reason, a large volcanic eruption has a cooling effect on the Earth generally, and this is reflected in a narrowing of tree rings at that time.

In 1883, for instance, the island volcano of Krakatoa, between Java and Sumatra, exploded and produced sounds that could be heard three thousand miles away. In 1815, Tambora, another volcano in Indonesia, had erupted and sent even more dust into the stratosphere than Krakatoa later did, producing "the year without a summer" in New England in 1816. The tree-ring pattern marks these eruptions, and other known cases, with narrowed rings to signify a cool summer. This means it may be possible to mark off ancient volcanic eruptions where we *don't* know the exact date. For instance, we know that Mount Saint Helens erupted about four thousand years ago, but tree ring data tell us that the exact date is very likely 2035 B.C.

Again, the island of Thera in the Aegean Sea exploded, historians think, about 1500 B.C., and that helped destroy an advanced civilization that flourished then in Crete, and plunged the eastern Mediterranean into a dark age. *But* tree-ring data seem to show that the date of the eruption was 1626 B.C., which is a little too early for the historians, who may have to do considerable altering of the history of the period if the tree-ring data hold up.

The study of the past, of course, can be of great use for the study of the future. If we learn more about the influence of volcanoes on climate, and of climatic patterns generally over the last eight thousand years, we may be able to make more sense out of climates we may expect in the future, and this could be of the greatest importance to us.

NOTE: *When this essay first appeared in June 1984, it was called "The Five-Thousand-Year Calendar." Fortunately, a kind letter from William J. Robinson of the Laboratory of Tree-Ring Research at the University of Arizona brought me up to date, and I made several improvements in the article before allowing it to appear here.*

8 THE THIRD SENSE

HUMAN BEINGS ARE BLESSED with three long-distance senses: seeing, hearing, and smelling. Of these, the longest-distance sense is seeing. We can see streams of photons crossing the vacuum of space and, as a result, we can make out the Andromeda galaxy with the unaided eye, even though it is 2,300,000 light-years away.

Hearing and smelling are confined to the atmosphere. If sounds are loud enough, they can be heard for many miles. As noted in the previous essay, the explosion of Krakatoa a century ago could be heard three thousand miles away; the vibrations of the atmosphere it produced were detectable by instruments all over the world.

Smelling, the third sense, is the least long-distance. A male moth can smell a female of his species a mile away, but human beings can't do that. In fact, we human beings make such steady use of the senses of seeing and hearing that we tend to ignore the messages that smelling brings us unless they are quite powerful or quite offensive, and we therefore do not realize how effective and delicate a sense it is.

Whereas seeing detects photons and hearing detects vibrations, smelling detects molecules. There are numerous molecules that affect the olfactory receptors of the nose, and it is estimated that we can detect up to a thousand different types of molecules, each with a different smell, and can then detect various combinations of these for vast additional numbers of detectable smells.

Animals, which depend to a large extent on their ability to smell, can distinguish individuals as easily by smell as we can by sight. Think of a bloodhound's ability in this respect. We can't very well duplicate the ability of the bloodhound, but we can do far more with the sense of smelling than we think we can. Tests have been reported that tend to show people can distinguish between the clean perspiration odors of male and female, for instance. Members of a family can identify the smells of clothing worn by other members of their family and can distinguish among them. Here's something else to consider. The instant a light is put out, you can no longer see it. Photons do not linger, but vanish at once. When a sound ceases, you may hear it die away in echoes and reverberation but in a short while it, too, is gone.

The molecules we detect by smell, however, *do* linger, sometimes for long times. Cooking smells hang around, as we all know. Our noses may become saturated so that we become unaware of them, but if we leave the house and then return later, we can still smell the breakfast bacon or the dinner cabbage. In the same way, bloodhounds can follow tracks by tracing the smell produced by people and objects that passed by but are no longer in the place where the hounds are sniffing.

To put it briefly, you cannot see or hear the past, but you can *smell* the past. It may be, in fact, for that reason that smells are so much more evocative and bring back the past so much more vividly than sights and sounds do. All in all, it is a shame, then, that human beings don't make better use of such a versatile sense. But if we don't, perhaps we will build machines that can. It is well known that roboticists are trying to devise robots that can see and hear. Apparently, there are also those who are trying to develop a mechanical smelling apparatus that might someday be fitted into a robot.

The trick is to devise gas-sensitive semiconductors that will react with one molecule or another. Ideally, each semiconductor will react with only a single type of molecule, and the electrical characteristics of the semiconductor will change as a result of contact with that one type. In this way, the mechanical nose will produce various electrical peaks signifying particular molecules, and will be able to distinguish between them and between various combinations of them.

By making the mechanical nose sensitive to certain odors, we can have a device that will detect smoke or escaping gases or pollutants of various sorts before the human nose does (or, if the human nose does, the odors will be mechanically detected even while we are in the process of not paying attention to our own sense). Then, too, a mechanical nose may improve the ability of a robot to do the cooking. It would be a great help to follow the progress of the meal by smell, and to be able to detect any sign of burning. Think how marvelous it would be for a robot to know *exactly* how much garlic to add.

Finally, if a mechanical nose is programmed to react precisely to the smell of a particular child, a robot possessed of such a nose would be able, bloodhound-fashion, to do more than merely keep the child within seeing or hearing range. If for some reason the child had drifted out to parts unknown while the robot's attention was distracted, the robot could still track the child down by the characteristic trail of molecules it left behind.

---

9 NO ACCOUNTING

Judging the taste of food and drink is a hard task indeed. There's only one way to do it—and that is to use human tastebuds. We've all seen the demonstrations on television in which people taste each of two or more competing products and then select the one that tastes the best. Nothing short of a human decision would be in the least convincing (and even that isn't totally persuasive, because it is always the sponsoring product that wins).

And yet there must be some sort of chemical or physical difference between two products; some distinction in homogeneity, feel, composition, additives, *something* that has an effect on taste. Otherwise, different brands or varieties of a particular foodstuff would taste precisely alike, and we know that isn't so. We all have preferences among brands of this or that. Since that is the case, there ought to be some way of measuring the differences that exist.

In 1982, for instance, a Californian technologist reported on the results of shining laser beams through various types of cola drinks in one set of experiments and shining them through various types of wines in another set. Some of the laser beam just passes through transparent liquids and comes out the other side as though the liquid weren't there. Some of it is absorbed by the liquid and is turned into heat. Finally, some of the laser beam passes near some particular molecule in the liquid, or near some very tiny undissolved particle, and is made to veer from its path. Some of it, in other words, is "scattered."

The amount of scattering differs with the angle at which the beam hits the surface of the liquid ("angle of incidence"). It also differs with the nature of the particles and molecules that are encountered. The amount of scattering can be recorded on film, and for each brand of cola or wine, there is a characteristic pattern of scattering as the angle of incidence changes. In other words, it should be possible to identify a brand by a scattering experiment. (It might even be possible to tell the exact vineyard from which a particular wine is derived or the exact bottling factory in which a particular soft drink originated.)

Presumably, as time goes on, other types of tests will be worked out to deal with opaque and solid materials. There is a whole battery of techniques now used by physicists and chemists to analyze materials — tests that involve bombardment by x rays, by speeding subatomic particles, by ultrasonic sound waves, and so on. One or more of these tests may be suitable for various types of foods.

Patterns of one sort or another might be worked up for each brand of peanut butter, let us say, and for each variety of apple, and for lettuce grown with the help of chemical fertilizers as opposed to those using organic fertilizers, and for chickens grown in different states and on different feeds and under different conditions, and so on. What's more, these patterns might be analyzed into a series of numbers, and those might be stored in computer memories.

Would such numbers tell us which brand or variety of item tastes better? Of course not! As we all know, there is no accounting for taste. If one brand *really* tasted better than another in an absolute sense, then everyone would buy that brand as soon as he or she encountered it, and every other brand would go out of business. In actual fact, when there are a number of different brands of a particular item in existence, every single one of them sells to some extent, and there will always be people who will insist, quite truthfully, that any particular one of them tastes the best.

Nevertheless, if market research showed one brand to be particularly popular, other brands might try to duplicate its pattern number. Or a brand might produce varieties with different pattern numbers and place them on sale in different parts of the country to see which variety sold best over all, then concentrate on that one.

Or perhaps market research might show that in one part of the country, pattern A sold well, while in another part, pattern B sold well. A concern marketing the brand might carefully adjust additives or method of manufacture to produce various patterns, which would be distributed accordingly. Or generic food items might be produced cheaply, with the pattern number recorded on the label, and customers might choose that item which most closely matched the more expensive brand they would normally buy. Or the household cook might learn the effect on the pattern number of a particular dish of each condiment, each spice, each variation in mixing, smoothing, slicing, heating. There might even be a home pattern analyzer that would evaluate the number as cooking proceeded, and the cook might labor to adjust matters until that number exactly matched the value that the family had learned to love.

Or—just perhaps—no amount of analysis will ever match the unmeasurable instinct of the skilled chef, for if there is no accounting for taste, there is also no accounting for creative genius.

10 DREAMS

DREAMS HAVE FASCINATED human beings from the dawn of recorded history, and very likely through the long ages of prehistory as well. There is something mysterious about those things you see and experience that belong entirely to you, that no one else shares, and that seem to have no necessary relationship to reality. It must have a relationship to *something;* if not to reality, then to something behind or above reality, perhaps.

In the Bible, Jacob dreams of angels traveling between heaven and earth and recognizes the place as holy ground (Genesis 28:11–22). Later on, Joseph has prophetic dreams that show he will someday rule his brothers and parents (Genesis 37:5–11). Eventually, in the most famous case of the sort, Joseph interprets the significance of Pharaoh's prophetic dreams (Genesis 41:1–36). The Book of Daniel and the Gospel according to Matthew also deal with dreams.

There, of course, we can suppose there may have been divine inspiration. In modern times, however, even where there is no question of that, there are still countless people who attach great significance to dreams and who feel that dreams are prophetic. It may be that almost everyone who reads this essay knows someone who dreamed something or other, and then, in a few days, found that something had happened in reality which "proved" that the dream had meant something all along.

There are numerous books that tell you the fortune-telling significance of anything you dream, and these (I am sure) find ready purchasers. There are undoubtedly people who swear by such books and who are guided by dreams in playing the numbers or the horse races or the stock market. And, every once in a while, I am certain someone makes money in this way.

Modern science puts no stock in dreams as a way of foretelling the future, any more than in the fall of cards, in the lines of your palm, or in the arrangement of tea leaves. Yet psychoanalysts believe that dreams tell something about unconscious thoughts, fears, and desires. With this in mind, they interpret dreams as zealously as ever Joseph or Daniel did, and here we have something much more likely to have significance. If you dream that your aunt Zenobia is sick, I don't for one minute believe that you have gained a glimpse into the future and that she will become ill. The dream may very well signify, however, that *you* have a problem because you have a perhaps unacknowledged desire to have something unpleasant happen to her.

What bothers me, however, even in the case of analysis, is not so much the reliability of the dream, as the reliability of the *report* of the dream. In my own case, I know I have the greatest difficulty in remembering dreams at all. Even when I do remember them, the nature of the events and the order of their occurrence are quite hazy and quickly grow hazier with time. Even if we sit right down at the moment of waking and write down everything we remember, how sure can we be that we have it entirely right?

After all, psychologists have shown over and over again that people are completely untrustworthy in their reporting of actual events that have happened in their very presence. A dozen people viewing the same event will disagree among themselves on every point, and not one of them is likely to get all the points correct. How much less likely are they to be correct in remembering something that has no objective reality, and concerning which no one can check them or correct them?

There is always a natural tendency to adjust your dreams to match later events. If you dream something—anything—about your aunt Zenobia and then two days later she falls and breaks her arm, you are sure to be tempted to remember that you dreamed she would break her arm. With each day you will remember more clearly that you had dreamed all the details correctly; you will gain prestige as a prophet; and there will be one more tale that "proves" precognition.

In the same way, a patient describing dreams to an analyst may not be able to resist augmenting their dramatic nature, and who knows how this might pervert the analytical process?

Back in March 1979 in my essay "Direct Contact" (see *Change!*, Houghton Mifflin Co., 1981), I wrote of the possibility of somehow analyzing brain-wave data by computer in such a way as to read thoughts. If so, might we not also learn to scan and record dream data by properly computerized electroencephalography?

Analysts might study such records the way internists study x-ray photographs or blood data, and when compared with what patients *say* they dreamed, the records might be illuminating indeed.

Another possibility would be the recording of dreams in large numbers in order to compare potential predictions with developing reality. I am certain that dreams are *not* a doorway into the future, and it would be nice to have the data demonstrating the conviction. However, I'll bet that wouldn't for a moment interfere with the sale of dream books or keep people from telling about the dream in which Aunt Zenobia had broken her arm and then, two days later —

NOTE: *This essay first appeared in November 1981. Four years later I had completely forgotten having written it (when one is as prolific as I am, that can happen) and proceeded to write another essay on dreams. At first, I was going to eliminate one or the other essay, but then I thought it might interest you to read both, since I approach the subject from rather different angles. The later (October 1985) article follows.*

11 PERCHANCE TO DREAM

WE ALL DREAM. There's no question about that. In fact, if we are deprived of dreaming time, we tend to grow psychotic, so dreams perform some essential function. The only trouble is that we don't know for sure what that function is.

In ancient times, dreams were thought to be visions of the future. In the Bible, we have Joseph's interpretation of Pharaoh's dream and Daniel's interpretation of Nebuchadnezzar's dream. And nowadays, there are any number of dream books you can buy that will tell you what every event in your dream portends. Many psychiatrists feel that dreams reveal unconscious thoughts and buried memories, so that a close analysis of dreams can explain a great deal that would remain mysterious if only conscious, wakeful thoughts were dealt with.

My own feeling, however, born of nothing more than what seems to me to be common sense, is that mammalian brains in general, and the human brain in particular, are so extraordinarily complicated that the process of gathering sense perceptions and processing them (to say nothing of the complexities of the abstract thought of which human beings are capable) inevitably snarls the brain's mechanism. During sleep, then, the brain "cleans house," unsnarling itself and getting rid of perceptions and thoughts that might represent the daily "garbage." If we must go without sleep for too long, that will kill us faster than going without food would, which is just a way of saying that a human brain's proper functioning is clearly more crucial than a human stomach's proper functioning.

Of course, who is to say that the process of cleaning out the garbage is totally efficient? Is it possible that occasional nuggets of useful material are dumped? It seems to me that this must be so. Useful thoughts may be drowned out by the busy chaos of the brain's daily work and then show up only during a dream as they are being discarded. If the dreamer can then remember that moment of garbage removal (if he wakes up as it happens, for instance), he may recognize its value.

The chemist Friedrich Kekule tells the story of having dreamed the molecular structure of benzene after devoting untold daytime thought, without result, to the problem. The physiologist Otto Loewi dreamed an experiment that would settle a problem of nerve chemistry that he was working on. He woke at 3:00 A.M. realizing that he had solved the problem, went back to sleep, and found he had forgotten the matter by morning. The next night he woke at 3:00 A.M., again with the problem solved, but this time he dressed, went to his laboratory, performed the experiment and, eventually, won a Nobel Prize for it.

The English poet Samuel Taylor Coleridge claims he dreamed his poem "Kubla Khan" and, on waking, began to write it down from memory as quickly as he could, until a person from

Porlock interrupted him on some mundane business. Afterward, he could no longer remember the rest of it and so was forced to leave the beautiful poem a mere fragment.

I was myself the beneficiary of these dream illuminations on at least two occasions. On April 3, 1973, I woke suddenly and told my wife that I had dreamed I was preparing an anthology of those science fiction stories I had read and loved back in the 1930s when I was a teenager. "How I wish I could really reread those stories," I said.

"Why not do the anthology in reality?" she said. Why not! I called Doubleday as soon as they were open for business that very morning, made the necessary arrangements, and on April 3, 1974, the anniversary of the dream, that very anthology was published under the title *Before the Golden Age*. In this case, the dream even seemed to have predicted the future. Of course, I *created* that future, but many prophecies are self-fulfilling.

A few years ago, I dreamed I had followed a person into a dining room and he had seemed to disappear. Since, even in my dreams, I don't accept anything illogical, I searched for him and finally found him. I admired the cleverness of the hiding place and said (in my dream), "What a terrific gimmick for a *Black Widowers* story." (The *Black Widowers* stories are a series of mysteries I write.) Once I woke up, I sat down and wrote the story, using the gimmick I had dreamed, and sold it for publication under the title "The Redhead."

Who knows how many of my good ideas get swept out with the garbage, for, unfortunately, I almost never remember my dreams. I can only think, wistfully, that if it were ever possible to record dreams as they take place and those dreams were then examined, the number of useful thoughts produced by creative people would be greatly increased and the world might be immeasurably better off in consequence.

I must admit, however, that I can't imagine how dreams can possibly be recorded. Could I *dream* a way, do you suppose?

NOTE: *Now that I've read both my "dream" pieces one after the other, I must admit I like the second one better, so I'm glad I forgot writing the first. (Some psychiatrists might say I deliberately forgot the first so that I could write the second.)*

12 STARTING POINT

THERE ARE TWO WAYS of looking at the matter of life in the universe: optimistically and pessimistically.

An optimist, thinking of all the stars in existence and all the planets that circle them, decides that there are billions upon billions of planets that are very much like Earth. He feels that life must have gotten a start on all of them, and that, on a certain percentage of them, intelligent life and advanced civilizations must have developed. There might be millions of them in our Galaxy alone. But then, with all those civilizations, many of them probably far more advanced than our own, there should have been space ships exploring the Galaxy for billions of years. Yet there is no real evidence that extraterrestrial civilizations have ever arrived here.

A pessimist, thinking of life, decides that it is so complex a phenomenon that it has only an infinitesimal chance of getting started *anywhere*. Life did get started here on Earth, but it may be that this planet is one of the very few places in the universe, if not the *only* place, where this rare phenomenon, life, actually reached the point of intelligence and a technological civilization. The difficulty with this view is that it isn't easy to argue that there is something special about Earth. Is there some key ingredient we possess that is found nowhere else? Is life the result of some near-impossible fortunate chance that somehow just managed to take place here?

Recently, a compromise view has arisen which we might call "opti-pessimism." Even though life is difficult to start, it may be that there is a full universe anyway. Essentially, the suggestion is that we ought to distinguish between the ability of a planet to *originate* life and the ability of a planet to maintain and evolve life *after* it has been originated. It may be that there are very few planets on which life can originate but very many on which it can be maintained.

Consider the Earth, on which life seems to have originated somehow. Once life started, however, as very simple microscopic cells, the rest was comparatively easy. Multiplication, variation, mutation, natural selection, allowed it to continue and evolve, finally reaching human intelligence and human civilization in the ordinary evolutionary process.

But now human beings are reaching out to other worlds. Twelve different human beings have stood on the Moon, and human instruments have landed not only on the Moon but on Mars and Venus, too, and are still there. Undoubtedly, in time to come (if we preserve civilization), human instruments, and perhaps human beings, too, will stand on still other worlds.

It is very easy to contaminate such other worlds with earthly bacteria in this way, if the world is in the least hospitable. We might do so carelessly, or we might even do so deliberately, just to get life started elsewhere. So far we haven't, because the Moon, Venus, and Mars are not really hospitable worlds. Life as we know it (without advanced technology) is flatly impossible on Venus, probably impossible on the Moon, and may be impossible on Mars.

But what if we journey to other stars or send instruments there? There, it could be, planets exist that are quite Earth-like, and even if no life had formed on them, microscopic Earth-life, falling in the alien ocean, might find chemicals it could live on and begin a new process of evolution. This, in a billion or 2 billion years, could produce an intelligent species that might be nothing like us at all, yet would be descended from a common stock. At some stage in the future, we may seed a million planets one way or another, and if, on a hundred of them, advanced civilizations develop (each one different), each of them may begin to seed still other planets.

Even if life could start only on a very few planets here and there, by now there may be numerous life-bearing planets. Even if we are the *only* example of life in the entire universe, then, provided we preserve our civilization, we may be the starting point for an eventually crowded universe. We need do nothing more, perhaps, than send out spore-filled capsules.

And maybe Earth is not the starting point. What if we are one of the seeded planets? Nobel Prize–winning scientist Francis Crick speculates that life was started on Earth a few billion years ago as a result of accidental or deliberate seeding from some far earlier civilization elsewhere. It's just speculation, of course, since there is no evidence for it, but if this is so, where, we might wonder, would the starting point of Earth-life have been, and what happened to that ancestral planet? Is it still there for us to find someday, and if we do find it, could we recognize it and know it for what it was? Or (since a few billion years is a long time even in the history of a planet), could it have long since perished? Will we ever know? Can we ever know?

13 STANDING UP

WHEN WE THINK of human beings compared to the different apes, there are three big differences that mark us off from them. One is, of course, the brain. Our brain is much larger than that of any ape. The second is the hand, which has a large opposable thumb, and which is therefore a particularly delicate and mobile instrument. The apes have thumbs which, by comparison, are small and feeble. The third difference is that we stand upright permanently and can do so gracefully and without difficulty, whereas the various apes walk on all fours at least part of the time and when they progress on their hind legs, as they do some of the time, they seem to do so uncomfortably and tend to shamble.

Which of the three differences is the most important one? It might seem to us, thinking casually, that surely our giant brain is *the* most important item. We could do nothing truly human without that brain. No other primates, not even the chimpanzees, have the mental equipment to make the kind of complicated sound modulations that produce true speech. And beyond speech, it is surely only our brain that makes it possible for us to think, to reason, to imagine—in short, to be human.

In second place, it might seem, would be the human hand— delicate and flexible. What an instrument with which to examine the universe, whether to grip something strongly or pick up something daintily and hold it before our eyes.

In third place would be our upright posture.

All this, though, is speculation, and if all we could consider were just ourselves and the apes, we might never be able to go beyond that. However, there are other organisms we can study. There are places in the rocks where scientists have discovered fossilized remnants of skulls, jawbones, teeth, thighbones, and other objects that are anywhere from a few hundred thousand to a few million years old. These fossils are noticeably different from the equivalent parts of human beings, but they are more nearly like those of human beings than those of apes. They are therefore said to be parts of organisms called "hominids" (from the Latin word for "man.") They are not "modern man"; they are not *Homo sapiens;* but they are more manlike than apelike.

A fossil skull of the earliest type of hominids, existing more than a million years ago, was discovered in South Africa in 1924. The discoverers called the organism to which the skull belonged *Australopithecus africanus* (from Greek words meaning "southern ape of Africa"). It wasn't an ape, however; it was a hominid. A number of such bones have since been discovered in Africa, and it is now known that there were several different types of such early hominids, which are grouped together as the "australopithecines."

In 1977, the American archeologist Donald Johanson discovered the oldest example of an australopithecine yet found. He discovered enough bones to represent about 40 percent of the entire skeleton, and since they are clearly the remains of a female, the name Lucy was somehow attached to the skeleton. Her scientific name is *Australopithecus afarensis*, and she is about 4 million years old.

Lucy, apparently a young adult, is only about three and a half feet tall. Her brain is small even for her size. She can't have been much brainier than a chimpanzee, and she must have been quite chimpanzee-like in facial appearance. The australopithecines, in fact, may have only been able to vocalize in chimpanzee-like fashion. We can't tell much about the hands, for hand bones don't seem to survive very well.

One thing, though, we can surely tell from the skeletons we have. The australopithecines, even ancient Lucy, walked fully upright, and did so just as easily as we do. From this, it seems that the change which marked the origin of all that was human was the matter of upright posture. Many animals sometimes stand up on their hind legs now and then. This raises their eyes and nose and allows them to see and smell food or enemies further. But the hominids did something other animals (even apes) didn't happen to do. They developed a unique, doubly curved spine that could support them upright indefinitely.

Once a hominid stood upright, its forelimbs were completely

freed from the task of body support; they were used only for manipulating and inspecting objects about them. Any change which made them more suitable for that purpose improved their ability to survive. It meant longer life and more young, who would inherit the better hands.

The more the hands could be used to handle and inspect, the more information flooded into the brain. Any change that happened to make the brain larger was therefore useful and prosurvival. That meant longer lives and more young who inherited the better brain. The size of the brain grew rapidly (by evolutionary standards) and the giant object we have in our skull is apparently the direct result of the ability of our ancestors 4 million years ago and more to stand up.

14 THE DNA "FINGERPRINT"

LATER IN THIS BOOK there is an essay entitled "Four Times Four Times Four—" in which I try to describe the unimaginably huge number of different DNA molecules that could potentially exist. It is possible for every living thing that ever existed on Earth, from viruses to sequoia trees, to have had DNA molecules, no two of which anywhere in any creature were exactly alike, and yet that would not even have scratched the surface of the vast number that could exist.

Shouldn't we suppose, then, that different species of organisms would have DNA molecules that were, in actual fact and not just in supposition, different from those of other species, even when the species are very similar? Indeed, this is so. Biochemists have only in recent years developed techniques for determining the exact structure of DNA molecules, but they have already discovered that characteristic DNA differences between the species do in fact exist. What's more, the greater the difference in appearance and structure between two species, the more marked is the difference in the structure of the DNA molecules that the two species possess.

Thus, the differences in DNA structure are measurable but not very great between chickens and ducks. The differences in DNA structure are considerably greater when either a chicken or a duck is compared with a human being. The differences are still greater when the DNA structure of any warm-blooded vertebrate is compared with that of an invertebrate, and greater yet when compared with that of a plant. This technique is still in its infancy, but by comparing DNA structures from species to species and placing them either nearer each other or farther away, according to the amount of difference that is found, a new way of determining evolutionary history arises. Till now, the details of evolution have arisen chiefly (but by no means entirely) from fossil evidence. The few measurements that biochemists have so far made using the new technique seem to go right along with the conclusions derived from fossil and other kinds of evidence, which strengthens the evolutionary details we have worked out.

In fact, DNA evidence may help us come to decisions in cases where fossil and other kinds of evidence are inconclusive. We don't have enough primate fossils to understand exactly how human beings and the great apes split away from a common ancestor. If we study DNA, however, it may be possible (for instance) to decide that the gibbons, which are least like us in DNA structure, split off first. The remainder of the evolving ape line may have split into two branches, from one of which the gorillas and chimpanzees evolved and from the other of which human beings and orangutans evolved. Or we may end up with another scheme. We must wait until all returns are in on the comparative structure of DNA.

It might even be possible to estimate how long it takes for changes in DNA to take place. From the amount of difference in DNA structure between two species, it might then be possible to decide just exactly how long ago they split away from some common stem. Preliminary figures have been offered in the case of the primates, for instance. Again, this evidence could be matched up with the totally different kinds of evidence obtained by measuring the age of the rocks in which particular fossils are found. The combination would then offer a strengthened picture of evolutionary development.

But is it only between species that DNA differences are to be found? Individuals of a particular species differ slightly among themselves; we can easily recognize particular human beings just by their facial differences, for instance. Mustn't this be a reflection of DNA differences? Geneticists at the University of Leicester in England have studied *related* human beings and have found that there are differences which tend to affect a tenth of one percent of the regions in the DNA molecule. That is enough to detect. Undoubtedly, if we were to study the DNA molecules in every human being on Earth, we would find that (except in the case of identical twins, perhaps) no individual human being would have a DNA structure which in every detail was exactly like the DNA structure of any other individual. In effect, you would have a DNA "fingerprint."

This would offer a new and very fundamental tool to follow the manner in which inherited diseases or a tendency toward disease is passed along from parents to children. It might also offer a delicate way of determining whether hospital mix-ups occur in the case of newborn babies, or of settling paternity disputes, or of reaching decisions in forensic problems generally (what a boon for detectives!). On a larger scale, scientists could in this way study whole populations, try to follow patterns of inbreeding and migration in the past, and so on. And we might acquire detailed information on the way in which DNA molecules combine, recombine, mutate, and so forth. The possibilities are enormous.

15 OUR EVOLVING BODY

THERE IS NOTHING CHANGELESS about ourselves, as we all know. Our hair turns gray or thin, or both, with time. Our waistline expands (or contracts, if we are grimly determined); we develop aches, or they disappear; and so on. No use going on with the litany.

But what about humanity as a whole? We are the crown and triumph of the evolutionary process; doesn't that mean the whole thing has come to a well-deserved halt and we can look forward to an indefinite future of that magnificent creation — the human brain and body? After all, the Stone Age people who produced the brilliant cave paintings in what is now Spain and France were as tall as we are, as handsome, as intelligent, and as artistic. And that was twenty-five thousand years ago.

Not so! The human body is in the process of transition. For instance, it has not yet, after all this time, completely adjusted to its two-legged gait. The spine tipped on end when we got on our hind legs and developed an S-curve in order to bear the body's weight in its new position. That's not good enough, so a large percentage of human beings suffer, on and off, from lower-back pains that we'd know nothing about if we were still on all fours. In addition, our sinuses slant upward rather than downward, now that we walk upright, and have trouble draining. Anyone with sinus trouble knows what that means. Then, too, we have an appendix we don't need. The appendix is very handy for some herbivores; a rabbit couldn't get along without one. The human diet, however, makes an appendix unnecessary, so it has dwindled away in the process of evolution, but not entirely. There's just enough of it still present to grow infected and inflamed on occasion and to kill us in agony if a surgeon doesn't intervene.

Our jaws have shrunk with the ages. We have hands now that stuff food into our mouths, so we don't need to be prognathous. We don't need jaws that stick out beyond our noses so that we can seize the food with our teeth directly. Our diet is pretty soft, thanks to cooking, for one thing, so we don't need enormous jaw muscles. However, the jaw has shrunk a little faster than our teeth have, so that thirty-two teeth are often a few too many for the jaw to handle. Wisdom teeth don't manage to appear till the late teens, are often impacted, and easily decay. The first step in orthodontics, which so many children need, is often to get rid of four teeth, one on each side, upper and lower. Perfectly good teeth, but no room!

Our brains have expanded amazingly, tripling in size in half a million years. Excellent, one would say, for who wants to be a chump? The newborn baby, however, has a large brain for an object only a few pounds in weight—in fact, its head is its largest part—and the adult female pelvis has to have room for that baby's head to pass through. That pelvis has barely kept pace, and this helps make childbirth the arduous process that it is.

Can we depend on the evolutionary process to see us through? Will it, in time, strengthen our spine, redesign our sinuses, shrivel our appendix to nothing, get rid of some of our teeth, and enlarge the female pelvis? Sure thing! The catch, however, lies in that phrase "in time." It can take a very long time. The latest suggestions in the continuing scientific investigation into evolutionary mechanisms indicate that particular species may go through long periods of many millions of years in which they are stable and show hardly any change. Then, over a relatively short period (a few hundred thousand years perhaps), they may show sudden startling changes under special, stressful conditions applied to relatively small numbers of them.

Thus, humanity underwent a rapid evolution in the last million years—but even "rapid" evolution isn't rapid in terms of human lifetimes. Consequently, we needn't expect the next few centuries to do anything at all in the way of ameliorating our anatomical and physiological problems; and they certainly won't help us adapt to the additional pressures we are placing on ourselves, such as pollution and unbearable crowding.

Ah, but we are human beings and we have brains. That has brought us to the point of developing the techniques of genetic engineering. Can't we look forward to a near future in which we can redesign human genes in such a way that we can arrange for babies to be born without appendixes or wisdom teeth and to develop larger pelvises for *their* babies, and so on?

Yes, we can, but there are two catches. First, we have to learn a great deal more about genes and gene combinations than we know. We have to learn techniques for locating and altering genes, and we have to know exactly which changes in the genes produce which changes in the body. Second, what about the side effects of changes that *appear* useful? For instance, if we reduce the number of our teeth, or their size, we save ourselves trouble, but we increase our dependence on soft food and further specialize our bodies. That may mean less ability to survive in the long run.

It will take lots of thought—and lots of hesitation.

16 THE ULTIMATE COMPLEXITY

IN ALL RESPECTS but one, the human being is not notably more complex than other living organisms. A lowly bacterial cell has the ability to carry through any chemical reaction we can manage and, in addition, a number of them that we cannot. Any plant cell can manufacture cellulose, an amazing substance which is the basis of wood—whose full gamut of properties we cannot duplicate with our cleverest synthetics. Yet our cells cannot form cellulose, nor can those of any other animal.

If we compare ourselves with other animals, then every organ we possess, except one, is totally indistinctive. Other animals are stronger, swifter, more sensitive, more sharply acute, more enduring, and so on endlessly. The one exception, of course, is our brain. We have a brain that can be considered superior to any nonhuman brain in existence. At least, it is the human brain *only* that has given rise to a technological civilization, and it is widely accepted that only the human brain is capable of deep, abstract thought.

The human brain is far larger than that of almost any other animal, including many animals that are far more massive than ourselves in total, such as moose and rhinoceroses. The human brain is not only enormous in size, but it must be extremely complex in order (to take an obvious example) for it to control the pattern of tiny, rapid motions of tongue, lips, and palate that make speech possible. In fact, it is reasonable to suppose that the human brain is the most complex organization of matter that is known to exist.

Considering that it is human folly and human madness that are the greatest dangers to human civilization, and even, perhaps, to human existence, it is extremely important that we learn to understand the human brain in detail. Surely we should learn to deal with its malfunctioning on at least the same level of understanding as we deal with the malfunctions of other human organs. And yet such is the complexity of the brain that it might seem almost hopeless to expect to understand it thoroughly. After all, how can the human brain expect to understand itself? One complexity can be understood only by a greater complexity; a brain by something greater than a brain, one might think.

Yet it is not a single human brain that is expected to do the job, but the many brains of the scientific community acting in cooperation, and many brains are more complex in total than a single one. Second, the brains do not work alone but have their tools, the most important of which is the computer—a tool that is being rapidly and constantly improved. Finally, we will be able to work our way up to our brain in stages, for there are other animal brains, simpler than our own and therefore easier to understand. They may offer us steppingstones to our own ultimate complexity. Particularly interesting in this respect are those few brains that are superhuman in size, even if not (perhaps) in complexity or ability; they may offer us the final step upward.

The average adult human being has a brain weighing about 50 ounces (a little over 3 pounds), but the male African elephant has a brain that weighs four or five times as much. The largest recorded elephant brain weighed 260 ounces. The largest brains of all are those of various species of cetaceans—the whales and dolphins. The largest animal that has ever lived is the blue whale, which may have a mass of up to 190 tons. It has a brain no larger than that of an elephant, however. A sperm whale, on the other hand, which may be only 70 tons in mass, can have a brain considerably larger than that of the elephant. The largest brain of any kind that has ever been recorded was a sperm whale brain weighing 325 ounces, or six and one-half times that of a human. The orc (sometimes called killer whale, but actually the largest living dolphin), which is only one-eighth as massive as the sperm whale, may have a brain weighing as much as 275 ounces. Even the common dolphin, which is no more massive than a human being, has a brain that is somewhat larger than the human brain.

We know that whales and dolphins are quite intelligent compared to other animals and that they have complex systems of communication. To be sure, they have no hands or other appendages with which to manipulate the environment, and they live in the sea, where it is impossible to develop a technology based on fire. Yet although they have not built a technological civilization, is that the only measure of intelligence? Might there not be other ways in which cetacean intelligence may be comparable to our own, and in which the orc may (just possibly) prove to be the most intelligent organism ever evolved? Certainly, if we develop techniques that will enable us to understand the workings of brains, it will be very important to use those techniques to try to understand the really large brains. And yet these large-brained animals are endangered, and it seems impossible to protect them. What a loss to humanity and to our hopes of understanding ourselves if they vanish!

17 HEAT WHERE IT BELONGS

THE ANCESTORS OF HUMANITY were tropical animals; at least, they lived in warm climates—and we're still tropical animals today. It wasn't until human beings learned to use fire and to wear clothes that they ventured into the merely temperate zones. We still use fire and wear clothes today to maintain a tropical temperature immediately next to our skin. If the fire goes out on a winter day and the clothes are ragged, the shivering and muttering that follows amply demonstrate our tropicality.

Not that fire was without its flaws to begin with. The earliest fires were troublesome. If they were built outdoors, they were likely to be put out by the rain. If they were built indoors to keep them safe from the weather, they quickly made the place uninhabitable with their smoke and smell. Before fires could be relied on as sources of heat, the chimney had to be invented. The chimney originated as a hole in the roof, but it eventually became the complicated brick structure with flues and whatnot that some of us know and all of us have read about. A stereotype of solid comfort is that of people crowding about the fireplace, talking, laughing, and holding tankards of ale while a roaring fire keeps them warm and happy.

People sigh for this only because they've never tried it. In actual fact, the roaring fire delivered most of its heat straight up the chimney. People crowded around it because only in its near neighborhood was there any warmth. The rest of the room was freezing. (In modern apartments and houses there may be working fireplaces, but a fire is lit only for its picturesqueness. No one would dream of shutting off the central heating system on a winter's day and letting the fire do the work. No one would dream of it twice, anyway.)

The first real improvement on the fireplace was arranged by that man for all seasons, Benjamin Franklin. It occurred to him that what was needed was an iron structure in the middle of the room, entirely enclosed and with a fire inside. The metal would heat up and radiate the heat in all directions. The smoke would travel up a bent stovepipe that led to the chimney, but the heat itself would not follow. Franklin built the first such "Franklin stove" in 1739 or 1740, and it worked just as well as he thought it would. The device spread and grew popular. The furnaces in the basements of modern homes are a kind of Franklin stove. (Franklin refused to patent the stove. He said that he enjoyed the inventions that other men had made and, therefore, he was willing to have others enjoy his inventions freely.)

A stove, and even an open fire, heats a room chiefly by means of the infrared radiation it emits. Infrared is made of waves like those of visible light, only longer. Infrared is produced more copiously than visible light is by ordinary household fires. The infrared waves are absorbed by the air, and by the walls and furniture of a room, and their energy goes to heating the objects that absorb them. The entire room warms up, and the cold surface of a person's skin, which also absorbs the infrared radiating from the stove and reradiating from the air and walls, grows warm, too. But why warm up the whole room when all we want to do is warm our body surface? There is a great deal of waste in warming everything in order to warm a particular something.

A Harvard physics professor, R. V. Pound, suggests that microwaves be used. These are just like infrared but have waves that are longer still. Microwaves that are a little over an inch long would not be readily absorbed by the air and walls of a room, but would be absorbed by the human skin and would heat up the water molecules in it. A person would feel just as warm then as she would if she were basking in the infrared radiated by the entire contents of a room. If this worked, it would represent the first real major jump in the efficiency of keeping human beings warm in winter (at least indoors) since the Franklin stove.

Of course, there has been talk of the possible dangers of microwave radiation; dangers which, in my opinion, have been much exaggerated. And even so, the microwaves that have been terrifying some people are the kind used in microwave ovens. These are five inches and more in length and would penetrate the entire body rather than just the skin. Then, too, ovens use microwaves in much greater concentration for the cooking of meals than would be needed for warming human beings. After all, we use high concentrations of infrared when we roast meat over an open fire or a gas flame or an electrically heated wire, but that doesn't mean we will roast a human being if we use lower concentrations of infrared to warm him. As in so many things, it's a matter of dosage.

Pound suggests that 60 watts of microwaves (equivalent to the power of a medium-size light bulb) would be enough to keep people comfortably warm in an ordinary living room. In these days of ever more expensive energy, letting the equivalent of a medium-size light bulb per room do the work of gallons and gallons of fuel oil is certainly an attractive idea.

18 GROWING THIRSTIER

CIVILIZATION MAY WELL BE thought to be the child of irrigation. It was agriculture that first made it possible for a group of people to build up a surplus of food. This meant that while some grew the food that fed everyone, others could follow other pursuits and become artisans, merchants, administrators, and priests—making up the complexity of a civilized society. But what agriculture requires more than anything else is water. There is no substitute for water. To be sure, water falls from heaven in the form of rain, but that rain falls spasmodically, unpredictably, whimsically. The fact that sometimes rain did not fall, so that the harvest failed, may have helped give human beings the notion of cruel and vengeful gods.

For an early civilization to develop mightily, something was needed that was more reliable than rain—and that was a river which contained a plentiful supply of fresh water whether the rain fell or not. The Nile, for instance, flows equably through a land that almost never sees rain and, once a year, it overflows and deposits fertile silt on both borders. Because of it, Egypt was the fabled land of plenty in ancient times. Similarly, the even earlier civilization of Sumeria developed along the Euphrates; the first Indian civilization along the Indus; the first Chinese civilization along the Hwang Ho.

The necessity of irrigation forced cooperation upon people. The water was no good to farmers as long as it stayed in the river, and it was impractical to try to collect it in pails and bring it to the farms. What was needed was a system of canals into which the river water would flow and through which the river water would travel to the farms of its own accord, along with dykes to fend off the water when, for any reason, the river level rose and threatened to flood the region. The canals had to be constantly dredged, the dykes constantly repaired, and that took the faithful labor of the entire community. What's more, it was not enough for a community to work hard at their irrigating procedure if some other community let their dykes fall apart and thus caused the entire country to be flooded—or let their canals silt up and then raided their neighbors for food when their own harvest failed.

The only solution was for whole river societies to cooperate, and that could best be brought about by a unified government, so that, sooner or later, some conqueror established an empire. That started a fashion that has continued ever since for varying reasons—but it began over matters of irrigation. And disputes over matters of irrigation have continued. There are still quarrels over water rights to rivers, not only in the perennially disturbed Middle East, but even in our own arid Southwest. And as world population has grown to levels undreamed of in the early centuries of agriculture, more and more water is required. We are drawing water from lakes and rivers in such quantities as to be threatening to leave them dry. We dig deeper and deeper for ground water, and the level continues to sink.

One of the problems, of course, is that for thousands of years human beings have simply assumed that the water supply was indefinitely large and would always be there. It seems inconceivable that we could run out of water, so we have made very little effort to economize. However, the world is growing ever thirstier and economy has become necessary. Fortunately, this can be accomplished.

Primitive water ditches have permeable beds, and much of

the water percolates downward and is lost before it reaches the crops. Some water that does reach the plants sinks deeply, past the roots, and runs slowly back to the river or lake it came from —or out to sea. And, of course, a great deal of irrigation water simply evaporates before it can be used. A better way is "drip irrigation," in which perforated plastic pipes are placed on or just below the soil surface. Water flowing through them trickles slowly out, more or less directly to the roots, with very little lost to evaporation. Less water is used than by older, more careless methods, and larger quantities are actually absorbed by the plants.

Irrigation can be computerized, too. The water flow can be adjusted, taking into account temperature level and wind velocities (on both of which the rate of evaporation depends), to say nothing of the quantity of moisture already in the soil. Leaks can be more easily detected, also. And if soluble fertilizer is added to the water, it, too, is fed to the plant more economically and efficiently. Israel, an arid land pioneering in this sort of irrigation, has done well with it. In the past fifteen years, it has managed to expand its irrigated acreage by 39 percent, while its water use has increased by only 13 percent. It's a step in the right direction, even though, in the final analysis, the only permanent cure is to stop the inexorable increase in the world's population.

19 BACK TO BASICS

IN PREHISTORIC TIMES, the chief tool-making material was stone. In fact, we refer to that period as the Stone Age. There were certain advantages to stone: There was a *lot* of it. It could be had almost anywhere just for the picking up. And it lasted indefinitely. The pyramids still stand, and the rocks of Stonehenge are still there.

But then, about five thousand years ago, people began using metal. It had advantages as well: whereas rock was brittle and had to be chipped into shape, metal was tough and could be beaten and bent into shape. Metal resisted a blow that would shatter rock, and metal held an edge when a stone edge would be blunted. However, metal was much rarer than rock. Metal was occasionally found as nuggets, but generally it had to be extracted from certain not very common rocks ("ores") by the use of heat. Finally, about 3,500 years ago, people found out how to extract iron from ores. Iron is a particularly common metal and is the cheapest one even today. Iron properly treated becomes steel, which is particularly hard and tough. However, iron and steel have a tendency to rust. About a hundred years ago, aluminum came into use (see essay 36). It is a light metal and can be made even stronger than iron, pound for pound. In addition, it is even more common than iron and won't rust. However, it holds on so tightly to the other atoms in its ores that a great deal of energy must be used to isolate it, so it is more expensive than iron.

In the twentieth century, plastics came into use. They are light materials that are "organic" (that is, built of the same atoms that are found in living organisms). Plastics can be as tough as metals; can be molded into shape; can be resistant to water and to deterioration such as rust; and can come in all sorts of compositions so as to have almost any kind of property we want. (Wood is a natural plastic that is still a common structural material and was probably used by prehistoric man even before stone was—but our forests are limited and are being destroyed at a dangerous rate (see essay 6), even though we have so many other structural materials to use.) *However,* plastics are usually derived from the molecules in oil and gas, and oil and gas aren't going to last forever. When oil is gone, plastics will be gone as well, for the most part. Then, too, plastics are inflammable and, in the process of burning, often liberate poisonous gases.

Well, then, is there anywhere else we can turn?

How about getting back to basics — to the rocks that human beings used before they developed the sophisticated way of life we call civilization? Rock remains far more common and cheaper than either metals or plastics. Unlike plastics, rock doesn't burn; and unlike metal, rock doesn't rust. Unfortunately, rock remains just as brittle now as it was during the Stone Age. What do we do about that?

Well, it might be possible to treat rock so that it would lose some of its brittleness. That would, of course, make it less cheap, but it would also become infinitely more useful; and, as in the case of metals long ago, the usefulness might more than make up for the expense. (This is all the more possible, as the expense has become more and more minimized.) For instance, different kinds of rocks can be combined and treated in such a way as to make the powdery substance wc call Portland cement. Water is then added, and molecules of water add on to the molecules in the powder, causing the powder to "set" into the hard, rocklike cement. As the cement dries, however, some of the water evaporates, leaving tiny holes behind. It is the presence of these holes that makes the cement brittle.

Scientists who work with cement have been developing ways of treating it during its preparation to make the holes formed by water evaporation much smaller than they would ordinarily be. The brittleness disappears, and you end up with cement that can be bent, that is springy, and that won't shatter on impact. Of course, it is important to search for a way of forming this sort of tough cement that would involve as little labor as possible and as little energy as possible. Scientists who work with materials are trying to find ways, for instance, of converting rocks into glassy materials without using the high temperatures required to form glass in the old-fashioned way. They are also trying to form ceramics and refractories (rocky materials that can be heated to very high temperatures and then cooled again without being changed in the process) in ways that consume little energy.

If all this works out, we may end up with relatively cheap stone that has all its own excellences plus some of those we associate with metals and plastics. We would then enter a High-Tech Stone Age that would mean a civilization far less wasteful of energy, far less concerned with preventing fire and rust, and far less subject to the disaster of dwindling resources.

NOTE: *After this essay first appeared in April 1985, a company making such nonporous rock sent me a sample. It is fascinating and makes an excellent paperweight.*

20 METALS FOR THE PICKING

WHEN THE UNITED STATES first gained its independence, it was a very loose aggregation of sovereign states. Congress didn't even have the power of taxation. It wasn't till 1787, eleven years after the great Declaration, that a Constitutional Convention set up a strong central government based on the federal principle, and this finally made the nation's survival possible. What impelled the states to get together, sacrifice some of their sovereignty, and accept compromise solutions? There were a variety of causes, but the issue that started the process going, in 1785, was how the use of the Potomac as a trade highway was to be apportioned between Maryland and Virginia. From that beginning, interstate meetings rapidly became wider and took up more general topics, until the happy culmination was the Constitution.

In the modern world, there are many people who believe that some global organization must be formed which can deal with global problems in a rational and pacific way; something stronger and less chaotic than the United Nations. How can such a thing come about? What is the "Potomac problem" that will start the ball rolling on a worldwide basis?

One obvious answer is the world's ocean. It belongs to no nation, and yet it is important to all as a highway of trade, as a source of food, and (increasingly in recent years) as a source of oil and minerals as well. For instance, the rivers of the world have been delivering traces of every kind of mineral to the ocean, slowly dissolving them out of the land they traverse. Despite this, the oceans contain comparatively little of these minerals in solution (and a good thing, too, or they would poison all life). The reason for this is that most of the minerals clump together into deposits which sink to the bottom of the ocean and gradually grow larger there.

After many millions of years of such growth, the ocean bottom is, in places, littered with "nodules" that form a rich ore of various metals. The Pacific Ocean, which is the oldest ocean as well as the largest, is particularly rich in these nodules. In places, especially in the North Pacific, the sea floor looks as though it is paved with cobblestones, so thickly spread are these potato-sized nodules. Occasionally, nodules have been dredged up and studied ever since the 1870s, and we have a good notion of what they contain. Most of them are very rich in iron and manganese. If this were all, it would be nothing to get excited about, since we have plenty of both. However, many of the nodules also contain small quantities of copper, nickel, cobalt, and other metals which are in short supply on land and which we could use. There are estimates that in the North Pacific, over an area half the size of the United States, some 15 or 20 trillion dollars worth of these metals abound just for the picking up. But all that wealth lies under a depth of 3 kilometers of ocean water, more or less, and the techniques required for dredging profitably at that depth can be worked out and developed only by a handful of industrialized nations—especially the United States.

Now comes a question. Is this untold mineral wealth to belong to the nations that dredge it up? If so, a few nations—the United States in particular—may become richer than ever, while the undeveloped nations will have to stand in line for what the industrially developed nations are willing to dole out at premium prices. And yet the ocean belongs to all, doesn't it?

This is the new Potomac problem. For ten years, the nations have been trying to work out a Law of the Sea. The less developed nations want the minerals distributed according to need, with the lion's share going to them, and they want the distribution supervised by a one-nation, one-vote system, which would give them a large majority. The industrialized nations want to profit from their own labors and don't want a system of "we do the work and you get the benefits." So far, in ten years, no successful compromise has been squeezed out.

There will be other problems, too. What happens once we establish a permanent station on the Moon and begin to exploit *its* minerals? There is a "Moon treaty" in the works, too, with precisely that question in mind, and negotiations there are hung up on the same intractable dilemma: The Moon is likely to be exploited by the United States primarily, but how much of the benefits can it expect to reap and what can other nations, particularly the undeveloped ones, do to earn their share?

As time goes on and human technology reaches down into the oceans and out into space, it may be best if all parties *do* compromise—if those that do the work get their reward but are satisfied with a moderate one. If so, it would not only solve the immediate problem but might also form the basis for further negotiation toward a global constitution, with incalculable benefits for all.

54

NOTE: *"Metals for the Picking" first appeared in June 1982, at which time it seemed possible that some headway was being made toward compromise in the realm of these international treaties. However, it all still hangs fire.*

21 STICK TO IT!

WHEN HUMAN BEINGS started to put things together, they needed some device to hold the things in place. They could bind a stone axhead to a wooden handle by means of windings of vine, for instance. Or they could just let gravity do the work. If you pile one very heavy rock on another—particularly if the two surfaces that meet fit each other quite well, so that there is no rocking—the two rocks are going to remain together. The pyramids built by the ancient Egyptians are made up of huge blocks of stone piled one on top of another. Nothing holds them together but their own weight.

But what if you need to hold light things together and tying them together isn't going to work? What if you want to put a thin piece of wood over a larger block in order to have a veneer? Or you want to put papyrus reeds together in order to form a sizable flat surface on which to write? In these cases, you would use glue. The ancient Egyptians used glue for such cases as long ago as 3000 B.C. Almost any sticky paste that would dry or set hard would do. A thin layer of flour paste or egg white or sticky materials extracted from animals' hooves, from fish, or from cheese could be smeared onto the materials one wished to bind. The materials could then be held together until the glue hardened, and that would be it.

To hold bulky pieces of wood together, wooden pegs could be

used. If the pegs were forced into holes a little smaller than they themselves were, the compressed wood fibers would hold them in place. Eventually, metal nails came to be used. The compressed fibers held nails just as well, and the nails were stronger than wooden pegs. Screws (harder to make than nails are) didn't come into use till about 1500. Nowadays, of course, we have any number of varieties of nails, screws, bolts, and rivets of all sizes, and they hold together houses, ships, automobiles, bridges, and everything else one can imagine.

So what about glue? Most people think of it as fit only for small jobs, for pasting paper, licking stamps and envelopes, constructing models made of balsa wood, fixing broken dishes, and so on. And yet nails, bolts, and rivets hold things together in a few specific places only. If a few of them break or twist off or rust through, and what remains isn't strong enough, objects will simply fall apart. Metal fasteners are not necessarily the ultimate.

Look at it this way. A lump of metal (or of any solid material) holds together because the atoms that make it up all touch each other in an orderly array. Strong interatomic forces hold the atoms together so that the lump is "all one piece." Now imagine two lumps of metal, each with a flat surface. Put them together, flat surface to flat, and they will not cling together. They will fall apart as soon as you stop pressing them together. That is because the "flat" surfaces aren't really flat. Under a microscope, they would appear quite uneven, and when the surfaces are put together, few atoms actually touch.

Suppose, though, you polished the surfaces so smoothly that they were truly flat, right down to the atomic level. When that happened, if you put those surfaces together, the two pieces would join and become quite literally all one piece. It wouldn't be very practical to do this. It would take too much time and effort, and it couldn't be done except as a laboratory demonstration. Riveting, or something like that, would be the practical answer to the problem.

But suppose you found a fluid that could spread out thinly

over one surface which you had polished moderately, so that it was reasonably clean and flat. The fluid would fill in all the holes, even microscopic ones, and then if you put the other flat surface upon the fluid, it would fill in all the holes in that one, too. Each flat surface would meet atom-to-atom with the fluid, and all parts of the fluid would meet atom-to-atom with themselves. If the fluid then hardened into a substance that was as strong and tough as the two objects being placed together, the whole would become all one piece without the necessity of polishing to atomic flatness.

The fluid is, of course, a glue, and if we could find a glue that was itself strong enough, it would hold objects together far more tightly and securely than nails, screws, bolts, or rivets would. The natural glues that people have been using for five thousand years aren't strong enough, but in recent years, chemists have been devising all kinds of synthetic resins that are far more efficient glues than anything ever seen before. Still more efficient glues may be created eventually, and if they are, the time may arrive when the steel beams of a skyscraper, for instance (and everything else of the sort: bridges, rails, machinery), would be all one piece. There would be no weak spots, no rivets or other fasteners to examine anxiously—only interatomic forces, which simply can't fail under ordinary usage. The world, as a result, would be safer than ever before.

---

22 HIGHER TOWERS

THE WORDS "CIVILIZATION" AND "CITIES" come from the same Latin root; and indeed, ever since civilization began, the world has become steadily more urbanized. Every century has witnessed the creation of more and, on the average, larger cities.

Prior to the industrial era, about 1 million people were all you could pack into a city. More than that, and you couldn't move food and water inward and wastes outward quickly enough to keep the city functioning. Nowadays, however, advanced technology has made it possible to pack 10 million people into a large city and twice that number into a large metropolitan area. Even under contemporary conditions, 10 million is rather more than can be decently managed, and yet there are predictions that cities two or three times as populous will exist within a generation.

There are hopes that a new age in electronic communications—new satellites, closed-circuit television, optical fibers, computerization, automation, robotics—will make it possible to control, administer, and maintain at a distance. People will not have to crowd together to do their work or to accumulate opportunities for culture. It may become possible for the first time in history to have people scatter, disperse, decentralize, and yet lose nothing. But how is this scattering to be done?

The American dream would appear to consist of a house in the suburbs with a nice bit of land about it; a two-car garage; a swimming pool; a well-maintained road or street at the end of the driveway; and a large shopping area within a mile's distance. Well, if every American family were given a house on an acre of land, the sprawl would take up one-fortieth of the total land area of the United States. And since comparatively few people would want their acre situated in the Alaskan tundra or the Dakota badlands or the Nevada sagebrush, much more than one-fortieth of the more convenient portions of the country would be taken up.

This would place enormous and devastating pressure on farmland and on woodlands. What's more, each house would have to be supplied with electricity, roads, sewers, water mains, and all the other appurtenances we have come to expect. The drain on resources and the strain on the environment would be insupportable. And if the whole world were to spread out, the situation would be even worse, for the population density of the world in general is one and one-half times that of the United States. Therefore, is there any reasonable way of decentralizing, since converting the world into suburban houses and lots wouldn't work?

The best way, of course, would be to reduce the Earth's population substantially, but, barring unbearable catastrophe, that will take a long time. Is there any way of decentralizing that offers some hope of success while we're waiting for a system of rational and humane population control to take hold?

One possibility is to centralize further and then decentralize the centralizations. Suppose we build skyscrapers a mile high, each of them capable of holding perhaps twenty-five hundred families. Each skyscraper would require an enormously complex system of water supply, waste and trash removal, electricity, gas, elevator system, heating oil, and so on — but in total, this might not be as much as would be required by twenty-five hundred separate houses each on an acre of land. The mile-high towers wouldn't have to crowd one on another, either. The population of Manhattan could be fitted into about 170 such towers, and if these were evenly spread out over the island, each would be about five or six city blocks away from its closest neighbors in any direction.

Each tower would be a sizable town in itself and might have all the community pride and esprit of a town. It would be largely self-contained, with its own stores, movie houses, libraries, gymnasiums, and so forth. Each tower would receive a great deal of sunlight, and the dark "canyons" of Manhattan would disappear. Much of the borough could become park land; parts could even be given over to vegetable farming. Transportation and communication among the towers would be comparatively simple, since roadways could be much wider than at present. The load of pedestrians might diminish, since personal travel between towers could be facilitated by enclosed bridgeways at different levels. This would also minimize the necessity of people moving up and down excessively and overloading the elevator system.

Of course, there would be disadvantages. The towers might encourage provincialism, with each person finding only enough patriotism within himself for his own tower. It would be all too easy to view inhabitants of other towers as dangerous foreigners. And what of the risk of fires, of "towering infernos"? What of the effects of hurricanes on such high structures? What perils would such huge, upward-stretching structures offer the nation's airlines?

What do *you* think?

---

23 NEVER AGAIN LOST

IN THE COMING AGE of communications, when modulated laser light will carry messages instead of radio waves, there will be so much room for different frequencies that every human being could be assigned a special frequency that would be as much his as his telephone number is today. Knowing a friend's frequency, we could tune in on it and reach him (thanks to communications satellites) wherever he might happen to be, provided he had his receiving terminal with him.

A person who could receive on a particular frequency that was all his own could also transmit on that frequency, and this might be very useful on certain occasions. We can imagine that there will exist in any police station in the crowded portion of the world, and in many outposts in sparsely inhabited areas, devices that will quickly scan the full range of frequencies and will be designed to zero in at once on anything within the range that is transmitting the equivalent of an SOS signal. (It will be analogous to the present radar set that scans the horizon for microwave signals.) Anyone, then, who is lost in a trackless wilderness will be able to set his transmitter to send out the necessary call-for-help signal on his own personal frequency and at a fixed intensity. The nearer receiving stations will detect that at once. If stations at several different points receive the signal and compare intensities, then it will be possible to determine the direction and distance of the point of origin of the signal and thus to pin down its location accurately. What's more, assuming that the personal frequency of each person would be on record in a sort of computerized super telephone book, the authorities would find out not only where the lost person was, but *who* he or she was. What it amounts to, then, is that no one would ever need fear being lost again.

Nor is it only a matter of being lost. You might be temporarily helpless. You might have had a heart attack or a bad fall; for one reason or another, you might not be able to move very well and there might be no one in sight to rally to your side. Your private SOS signal would then bring help. One can also see its potential usefulness in connection with children. Youngsters might well be outfitted with a continually active transmitter that had a range of a few city blocks. If a child was unaccountably absent, an anxious parent could tune in to the child's frequency and scour the neighborhood on the old hot-or-cold principle: the stronger the signal, the closer the child.

It would be a new variation of an old theme. Among animals, a lost young one will mew or bleat or squawk and the parent will recognize its own young one's sound among many others and zero in on it at once. A human parent will certainly respond to its own child's voice, but the identification could be made with greater certainty and precision, and at longer range, using a signal faint enough to be imperceptible to ordinary senses and therefore undisturbing to those not concerned.

To be sure, there is bound to be a clash of purposes here. A parent might well want to continue the beeping transmitter in full operation as the child entered his or her teenage years, and to make it longer range, too. In that way, the old question "Do you know where your child is now?" would be answered with a firm "Yes! Just as soon as I turn on my receiver and scanner and read the settings."

Suppose the transmitter was a micro-unit inserted under the skin; that it required a new energy source only once in a long while; and that any tampering set off an alarm. Kidnapping might become impossible. It might well be, however, that as the youngster grew older, he or she would not want the parents to know where he or she was every minute of the time. The older the youngster, the more this would be so. We can be sure of that.

There will be other conflicts of this sort: the desire to know on one side, the desire to stay private on the other. What about a woman who wanted to know where her husband really was when he said he was working late at the office? (Or, in these days of sexual equivalence, a man who was wondering about his wife working late at the office?) What about a curious neighbor who knew your frequency and wanted to know where you and your wife were heading as you stepped out of your house or apartment dressed in your best? What about the government that decided it couldn't feel really secure unless it knew exactly where every one of its citizens was at every moment of the day, and with whom he or she was huddling, and when and where—just in case dissent and sedition were being plotted? In short, if you think you have a problem about privacy now, just wait until the time comes when you may not be allowed to be lost.

NOTE: *This essay appeared in June 1981. The following essay, with almost the same title, appeared in March 1985. When I discovered the near identity in titles (too late, of course), I almost panicked at the thought I had written the same essay twice—but I hadn't. As you will see, the next essay is quite different from this one.*

24 NEVER GET LOST

ON LAND, WE CAN OBSERVE where we are and where we're going by noting landmarks and signs. At sea, on the other hand, there are no landmarks and no signs. The ancients learned how to get ideas as to the direction in which they were traveling by observing the position of the stars and the sun, but such a system was of no use on cloudy days. Even when the skies are clear, you need good instruments, including a good clock, in order to determine longitude accurately and thus know your precise position on the globe. These instruments didn't become available till the end of the 1700s.

Nowadays, ships can make use of signals from navigational satellites wheeling around the Earth and can determine their positions to within a few feet. They can do this at any time, night or day, cloudy or clear. As a result, it is now much easier to get lost on land than at sea. On land, what may be landmarks to the locals can well be unfamiliar to the traveler, and road signs may be absent, insufficient, or even misleading.

And yet navigational satellites can, in principle, be used on land as well as at sea. Why shouldn't land vehicles—even the family automobile—be equipped with a device that would detect signals from various satellites and translate them automatically into latitude and longitude? If the car were equipped with a screen that gave you your latitude and longitude at every moment, then you would already have useful information. Suppose you were in the northern and western hemispheres, as you would be anywhere in North America. In that case, if latitude increased as you traveled, you would be going north; if it decreased, you would be going south. If longitude increased, you would be going west; if it decreased, you would be going east. If both were increasing, you would be going northwest; if both were decreasing, you would be going southeast; and so on.

If you happened to know the precise latitude and longitude of your destination, you would be helped further. (And why would you not know them? Travelers who learned to make routine use of latitude and longitude would automatically get the figures for their destination, just as now they would get its number and street or its position along a highway.) Then, as you traveled, you would routinely observe your position and travel in such a way as to cause your latitude and longitude to change in the direction of your destination. When you had made the two sets of latitudes and longitudes identical, you would have arrived. Under such conditions, you would never get lost en route.

But watching the figures could be tedious, and trying to move in the right direction over roads that turned this way and that (to say nothing of the possibility of detours or of the absence of a cross street at a time when you had calculated you ought to make a right turn) could induce petulance and headaches. What you also would need is a map.

It is quite possible to have maps on cassette tapes and have them reproduced on a television screen. (Such maps are already reported to be in production). They could be prepared for special regions — for particular cities, for instance, or counties. They could be magnified to show particular portions in great detail, including names of streets, or reduced to give a general idea of the road network of a region. According to one scheme, the position of the moving vehicle can be shown always in the center of the screen, while the map moves and turns as you travel. The destination bears another mark, and at any time you can see how far away it is, in what direction, and what route you must take to get there. My own feeling is that it would be inconvenient to have the position of one's car presented on the map as unmoving, since that doesn't match reality. It would be better to have the map motionless and the symbol moving that represents one's car. Still, drivers might get used to the unmoving-car system, just as they have become accustomed to watching activity to the rear of their car through a mirror rather than directly.

Of course, making use of navigational satellites on land could be complicated, since there are competing radio signals of all kinds that might interfere, especially in cities. An alternative scheme would be to make use of dead reckoning. You would begin by knowing your position on the map and you would fix the car's symbol there. After that, as you drove, special sensors on your wheel would measure the distance you traveled, while a compass would note the direction. The mark of your car on the map would move accordingly (or the map would move), and you would always know where you were without having to worry about satellites.

One or the other system (probably both) could conceivably come into use within a few years, and after that, we will never have to ask, "Where are we?" We will *know*.

25 SHRINKING THE MICROCHIP

PEOPLE MY AGE can remember the first electronic computers, which were as large and as slow as trucks and used even more energy than trucks did. Computers have been shrinking ever since, however. Vacuum tubes gave way to transistors, which shrank rapidly. Individual transistors gave way to unified circuits, and these, too, shrank and shrank until we had the microchip. This is a small square of silicon on which the etched circuits are so tiny that they have to be viewed under a strong lens.

The result is that we now have pocket computers that cost very little, that run on small batteries or on exposure to a light bulb, and that can do more things thousands of times faster than the first computers could. It makes one wonder what can be done for an encore. How can one possibly devise computer components smaller than the microchip? We might dismiss the possibility out of hand were it not that there already exist computers with components far smaller than the microchip. Such computers have existed for a long, long time. The most advanced variety of them is referred to as the human brain.

The human brain contains 10 billion nerve cells and about 90 billion subsidiary cells. Each one of these cells is, in turn, made up of elaborate systems of billions of molecules, including, in particular, protein molecules. Even the largest molecules are extremely tiny compared to even the smallest microchip. Might we someday build computers with molecules serving as the basic components? With molecules storing and releasing data and carrying through computations? Scientists are already speculating on this possibility.

Suppose one synthesized molecules hundreds of atoms long (large for a molecule, tiny in comparison to microchips). Properly designed, such molecules could exist in two very similar configurations. A tiny pulse of energy striking one end of a molecule might travel its length, changing Configuration 1 into Configuration 2. A tiny pulse striking the other end might travel back and restore Configuration 1. Such molecules would act like switches, in other words, just as a vacuum tube or a tiny transistor would, except that a molecular switch could just barely be made out under an electron microscope. Pulses of energy running along the length of a molecule and changing the configuration are called "solitons." The phenomenon has not yet actually been detected, but theoreticians seem to be increasingly of the opinion that they can exist.

Then, too, there is the possibility of using carefully designed protein molecules. A protein molecule is built up of numbers of smaller molecules, called amino acids, that are strung together like pearls on a necklace. There may be hundreds or even thousands of amino acids in a single protein molecule, and they come in about twenty or so varieties. Each variety has a different "side chain." Some side chains are large, some small, some have a positive electric charge, some have a negative electric charge, and some have no charge at all.

Every different string of amino acids folds up in a different way and produces a protein molecule having a characteristic shape and a characteristic pattern of electric charges upon the surface. Even a slight change in the order of amino acids will produce a different protein, so the total number of possible different protein molecules is far, far greater than the number of atoms in the universe.

Protein molecules can usually exist in different conformations and can easily change from one conformation to another. In this way, they can serve as switches or as memory-and-recall devices; in fact, they should have the ability to do anything molecules can do in the brain. In the future, it may be possible to design proteins of specific shapes to perform different functions in computers. This may be done by designing appropriate genes and inserting them into bacterial cells. The bacteria will then proceed to produce quantities of the desired proteins. We can visualize the computer technicians of the future painstakingly supervising the growth of thousands of different bacterial cultures.

With combinations of different proteins serving as the "micro-microchips" of the future, we would begin to approach the creation of computers that would be no larger than the human brain yet would be capable of feats comparable to those of the human brain.

Such "protein computers" won't necessarily be identical to the brain, for they will probably be designed to deal with specific problems and to demonstrate particular types of behavior, but they will finally represent true embodiments of "artificial intelligence." And once we have a molecular computer as compact and as complex as the human brain, we will be able to fill the cranium of a man-sized, man-shaped robot and have it do the kinds of things I have been writing about in my robot stories for the past forty-five years.

---

26 SPELL THAT WORD!

SEVERAL MONTHS AGO, I received a word processor and was scared into paralysis by it. I mentioned this fact in an essay and, as a result, received a number of letters from concerned readers. Let me assure you all that I have managed to master the device and that I am sitting at it right now, composing this very essay. And since I notice that writers who discuss their word processors always insist on naming which of the many varieties they use, I may as well tell you that I own a TRS-80 Model II Microcomputer with a Scripsit™ program, and I am completely satisfied with it.

I will tell you that this is an enormous relief to me, for I do not have a high opinion of my manual dexterity, ordinarily. When people find out that I am using a word processor, they assume that my rate of production is greatly increased—but it isn't. You see, I was working at top speed at my typewriter. I type very quickly (ninety words a minute), and my thought processes keep up with my flying fingers. What's more, I revise very little. As a result, I turn out a complete article so quickly that there is very little room for acceleration by a word processor. It is the heavy revisers that generally gain speed, for it is much easier to pull apart an essay and put it together again on the television screen than on paper.

Why do *I* use a word processor, then? Because it enables me to turn in letter-perfect copy. My typing is fast because I don't bother with accuracy. My essays are therefore littered with typos, which I either correct (rather messily) or leave for the editor to correct. On the screen, however, the insertion, deletion, or replacement of a letter or two can be carried out quickly, easily, and without leaving a mark.

And that has gotten me to thinking about the matter of spelling. There is no language without flaws, and one major flaw in the English language is its spelling, which is simply ludicrously irregular. "Sure" is pronounced "shoor," but if you spell it "shoor," you are tabbed as illiterate. If you take to heart the spelling of "sure" and spell "shoot" as "sute," you are illiterate again. And there are all kinds of games you can play with "ough," which is pronounced differently in "through," "though," "cough," "rough," "hiccough," and "lough."

Nor can any plan for spelling reform ever be carried through. It is perfectly possible to have a language in which words are pronounced *exactly* as spelled, with no uncertainty attached (I understand that Spanish comes close to being an example of such a language), but no attempt to do this for English seems to have any chance of success. People have invested so much mental effort in learning how to spell according to our crazy system that any attempt to change that system and force a relearning is resisted tooth and nail. Yet if spelling *were* rationalized, true literacy would be greatly advanced. It would be much easier to learn to read and write, and there would be much more chance of making English an effective world language.

Might computers serve as a key to reform? At the present moment, I can change any spelling with childlike ease. If I write "speling," I can insert an "l" with almost no trouble, or I can remove it, if I write "spellling." I don't even have to look for misspelled words because I can make use of a special "dictionary" if I want to which will unfailingly spot and highlight any word that is not in it thanks to misspelling.

Suppose some central Bureau for Spelling Reform were to introduce new rules now and then. For instance, suppose we stopped using just "s" when we meant "sh," so that "sugar" and "sure" became "shugar" and "shure." It would be no skin off *our* nose. If we had become a society in which virtually everyone word processed instead of wrote, we simply would adjust our dictionary. Naturally, we would constantly be placing "sugar" on the screen (or "through" instead of "thru") out of sheer habit. Our dictionary, however, would tirelessly detect the error and nudge us, and we would correct it with hardly any trouble, until finally we had established the new habit. And young people, as they learned to handle a word processor, would have the new spellings from the start.

In fact, I look forward to a future time when a computer dictionary will contain common misspellings and corrections. For each misspelling, you will simply push a "Correct" button, and the correct spelling will automatically replace the error. Your only job will be to make sure the correction is the word you really want, and to pay attention to the error in order to encourage yourself to avoid repeating it the next time. We can even —I hope—look forward to a time when the word processor will interpret speech. You will speak and the words will come out phonetically (and, therefore, correctly, if spelling is made rational). And then we'll never have to spell again.

27 THE ELECTRONIC MAIL

ONE OF THE CLASSIC mystery stories is G. K. Chesterton's *The Invisible Man.* The invisible man wasn't physically invisible, of course; he was simply disregarded, so that no one noticed his comings and goings. He was the mailman. To me, this story never rang true, because I always watch for the mailman and would rather disregard half the world than miss him. I can't get my mail unless he hands it to me, and he takes the trouble to tie it up into a neat bundle for me.

Perhaps, though, Chesterton was merely ahead of his time, for the mailman, while not yet ready to disappear altogether, is certainly going to impinge on us less, because what we called when we were in school the "friendly letter" and the "business letter" may change their form completely.

Why write or type something on a piece of paper and then deposit that paper in a slot somewhere so that it will be physically carried, one way or another, for a distance of anywhere from one mile to ten thousand miles and be delivered to its destination in any time between one day and one week? It is not the paper itself that is of interest to anyone, only the words upon it, and those words can be carried electronically at the speed of light.

Suppose you had a computer terminal in your home into which you could slide your communications software whenever you wished to. Perhaps you could do so merely by pressing an appropriate button. You would then recite your letter and the words would appear on a small screen. You would correct it verbally, changing a phrase here and there, correcting a misspelling where you enunciated in slovenly fashion or where it proved difficult to distinguish between words such as "blue" and "blew," and eventually you would be satisfied.

You would then press another button, and the computer would go to work. It would address the letter, if the address were in its memory store. (If not, you would have to dictate the address and add it to the memory store if you expected to be corresponding more or less regularly with the addressee.) Included in the address would be a zip code elaborate enough to dictate the exact route that must be taken to the very building or apartment marking the destination, and then off would go the words, streaking along optical fibers at the speed of light. Indeed, it would be a form of light, a laser beam, that would carry your message.

Within a second after you were satisfied with the letter and pressed the "Transmit" button, the letter would have reached its destination. In another minute or so, it would have been converted from electrical impulses into printed letters on a sheet (or sheets) of paper, and those would be deposited in the letters-received basket. Naturally, a copy of the letter would be preserved for you even as it was being sent out. Those letters of any importance at all you would encode and store, for a period of time at least, in the computer's capacious memory. You yourself would no longer have to wait for the mailman to deliver your first-class mail. You would merely wait for the familiar ring or click that would tell you a letter had arrived. If you were away on an errand, a social engagement, or a trip, one of your first acts on returning home would be to check the mail basket.

Would even this whole process be necessary? If you were communicating electronically, wouldn't it be better simply to get in touch with the person image to image or computer to computer and engage in a spoken dialogue? In many cases, this would indeed be preferable. If, however, you had a message of moment to deliver, you might not wish to depend on the uncertainties of speech. You might want your words solidified on the screen so that you could adjust them and get them into the form that satisfied you and *then* send it. Or you might be interested in putting what you had to say on permanent file so it could be referred to in case of need. Even when none of this was so, the person to whom you wished to speak might not be available at that moment and a written message might then best answer your needs.

So much for individual apartments. In places of business, it might well be totally uneconomical to try to route pieces of mail to individuals. Letters to anyone in the establishment would all arrive in the same basket and would be sorted out robotically, perhaps, and then distributed to the proper individuals by means of a computerized mail cart.

And in all this, would the mailman really disappear? No, not entirely. It would probably remain uneconomical to electronically transmit bulky printed material such as pamphlets, magazines, newspapers, or catalogues. They would still be carried by the mailman, as would packages containing objects other than printed material. But with only such material to handle, the mailman might be more disregarded than ever and would be more truly, in Chesterton's sense, the invisible man.

28 BEYOND PAPER

WHEN WRITING WAS FIRST invented in Sumeria five thousand years ago, signs were incised in clay, which was then baked. The result was permanent (numbers of such baked-clay inscriptions still exist today) but *heavy*.

The Egyptians discovered how to make sheets of thin, light

material from the pith of a reed called papyrus, and brushed ink upon it. It was a paperlike surface (in fact, the word "paper" comes from "papyrus"), which was rather fragile but light and convenient. Egypt cautiously maintained a virtual monopoly on papyrus, however, and the reed, through overuse, became rare. The west Asian kingdom of Pergamum, therefore, in desperation, developed the use of "parchment" (a corruption of "Pergamum"). This was manufactured from animal skin and was far more permanent than papyrus but was also considerably more expensive.

Then, about A.D. 105, a Chinese eunuch, Tsai Lun, is supposed to have made one of the key inventions of human technology. He discovered how to make a papyrus-like substance, not from rare reeds, but from universally available plant products such as bark, hemp, and rags. This is what we now call paper. The secret of paper manufacture gradually spread from China throughout the entire world. It was this substance, cheap and available, that made mass literacy practical. The invention of printing in 1454 would have done no good if paper hadn't been available on which to print the floods of books that became possible.

The further development of even cheaper paper from wood pulp allowed a previously unimaginable proliferation of books, magazines, and newspapers. Indeed, the world, particularly the industrialized portion of it, is being buried under a growing mountain of paper. Where would we all be without the myriad of directions, forms, reports, memos, and so on—all of it print on paper. Think of all the blank sheets of paper each of us has at work and in the house to scribble on, to write notes on, to write formal expositions or friendly letters on—to say nothing of paper used for envelopes, wrapping, pasting, and various other purposes.

I remember reading a science fiction story about thirty years ago in which a scientist discovered a way of irradiating a region in such a way as to cause all the paper in it to oxidize more quickly than was normal, so that it all rapidly yellowed, grew brittle, flaked, and was gone. The United States used the technique on a certain enemy country which, in no time, found that its governmental machinery and its people's lives ground to a halt. Without paper, records could not be kept, and communication stopped. It was a thoroughly convincing story of a bloodless victory. (Unfortunately, libraries must deal with books whose paper is subject to this process—not in days, but in decades— and they are trying desperately to prevent the loss of priceless items.)

But now, for the first time since its discovery nearly two thousand years ago, paper is in danger of replacement, thanks to the new technologies. You might think that our high-tech world has only increased paper usage. After all, it is estimated that 600 million pages of computer printouts and 235 million photocopies are produced *each day* in the United States. But wait—

This essay is being recorded on a floppy disk that can hold about 125 pages of material altogether. Four such disks could hold my next novel, and they would be much lighter and less bulky than the ream of paper it would ordinarily take to hold it. If my publisher were properly equipped, he could use the disks directly and produce a book without ever using paper in the intermediate stages. The book itself could be placed on floppy disks and scanned on a screen instead of being read from the customary paper pages.

Of course, 125 pages to the disk isn't really very much. Computer technologists are constantly working on schemes to squeeze more and more words onto a floppy disk. It will probably not be long before it will be possible to make use of tightly focused laser beams to record the equivalent of 100,000 pages onto a single disk. In that case, we could squeeze all of the *Encyclopaedia Britannica* onto such a disk, with lots of room left over.

We can imagine a row of such disks, and not a huge row

either, with all of the Library of Congress recorded upon it. All of it could be classified and indexed appropriately so that anyone with a proper computer could call up any of it. A large fraction of communication transmission and storage could shift from paper to disks, and the space required for such matters would undergo an extreme miniaturization. This procedure would also solve the problem of crumbling, decaying books. Still, suppose a disk were stolen or damaged. How much worse that would be than if a book were! Or suppose something happened to disrupt our electrical technology, putting computers out of action for a period of time. At a bound, communication would come to a dead stop, as in the science fiction story I described—only worse.

---

29 SILENCE!

FOR THOUSANDS OF YEARS, writing must have seemed a self-absorbed, essentially quiet occupation. The ancient Sumerians, who invented writing, stamped styluses into soft clay. The ancient Egyptians and Chinese used brushes. There was, of course, the noisy task of chipping stone inscriptions, but they were monumental and comparatively rare.

To be sure, writing wasn't entirely quiet, even at its best. (What is?) Suppose we think of monks deeply absorbed in copying some old manuscripts or Gibbon writing his *Decline and Fall* or Dickens writing *Nicholas Nickleby*. Undoubtedly, they were all using goose quills, properly shaped and split, as pens. And, just as certainly, those quills scraped across the paper. In fact, the word "scribble" is an imitation of the sound that pens make. The dictionary says "scribble" is from "scribere," which is Latin for "to write," but I'm convinced that the Latin word deliberately imitates the sound. Goose quills were replaced, eventually, by the more durable steel pens, then by the longer lasting fountain pens, and finally by the very convenient ballpoints, and writing grew a little quieter with each step. A good ballpoint hardly makes any sound at all as it glides across the paper.

But who writes these days? I mean literally—with a pen and paper. The practice has declined enormously since Christopher L. Sholes invented the typewriter in 1867. Ever since then, more and more writing has been done by such machines. They are faster, they produce absolutely legible copy, and they are infinitely more convenient. I can't imagine myself writing my stuff with a quill pen now that I've been typewriting for nearly half a century.

There is a catch, however. The typewriter slams a key, incised with a mirror-image letter, against an inked ribbon, which in turn slams against a piece of paper. Each letter is perfectly formed and correct (assuming the correct key is struck) but appears only with a loud noise. Bang! Bang! Bang! goes the typewriter, a couple of times per second, to say nothing of the carriage sliding back and forth, little warning bells ringing, and so on.

The modern office has become a bedlam.

One gets used to it, of course. My noisy typewriter never bothers me. Even someone else's doesn't. Nevertheless, who knows the long-term psychological effects on the writer and on his writing? There was some worrying about this, and for a while, "noiseless typewriters" were all the rage. I even tried one, but I didn't like it. The keys didn't hit the paper hard enough, and the letters were uneven and too light. Nor were they really noiseless, either.

Besides, there was another kind of convenience that seemed more important—the removal of the need to slam the keys (a wearying process, take it from me). The electric typewriter became common, making it necessary only to touch the keys; electromagnetic forces do the rest. But electric typewriters were even noisier than hand typewriters, since electromagnetic forces never grew tired and since the harder the contact, the darker and more even the lettering. Electric typewriters also went faster, on the whole.

Finally, there came the word processor, which also requires merely touching the keys; but then, electronic forces simply produce the letters on a television screen without any forceful slamming. My word processor works with only a slight chuckle, and I can use it freely in the living room, whereas my electric typewriter must remain hidden in the recesses of my office. *But,* there comes a time when my screen letters must be printed out, and for that I use my printer. And then again, keys slam against paper, and do so four times faster than I could possibly manage with my own fingers. It will take me four minutes or so of nearly steady br-r-r-m to have this essay printed out.

So are we stuck with noise no matter what we do? Perhaps not. There are suggestions for methods of printing other than the effects of sudden pressure. It would seem possible to devise practical ways of squirting tiny bits of ink onto the paper so that the ink would dry in dots that would form letters. Or else lasers could be used to score invisible letters on a cylinder, and the ink would adhere only to those scored portions. If the cylinder then turned against a sheet of paper, an entire page might appear almost at once. In either case, the process of making print appear would be virtually soundless. We'd be back in the silence of the medieval cloister but with all the advantages of the speed and reproducibility of high-tech printing.

30 IMPROVING THE ODDS

There is something awesome about the manner in which statistics permeate professional sports, and I suppose that nowhere is this quite as true as in baseball. Every major league game of the twentieth century (and, I think, of a good part of the nineteenth century as well) is permanently engraved in the record books—every player, every hit, every strike, every base on balls, every catch, every error.

The old records live for decades and serve as marks to shoot at. Sometimes they fall. Babe Ruth's records for most homers in a season and most homers in a career both finally fell (to the indignation of many old-timers). Pete Rose has now broken Ty Cobb's record for most hits in a career, and DiMaggio's record for a hitting streak in most consecutive games is always there to inspire every rookie who seeks immortality.

The statistics, however, comprise more than a compilation of most this and highest that and longest something else. Given any player, you can find out how well he hits against left-handed pitchers as compared to right-handed ones, or how nearly error-free he is on Tuesdays as compared to Wednesdays, or (if he's a pitcher) how many bases on balls he will give up on warm, cloudy days as compared to cool, sunny ones. At least, this can be done in theory—if you go through the record books carefully enough and comb out the necessary details and do the required calculating. The trouble is that until comparatively recently it would have been so tedious and so time-consuming that no one would have dreamt of doing such a thing.

It might be useful, you understand. If you knew enough details about each player on your team—details about his performance under different conditions or when facing different types of opposition, about the exact situation in which he is most likely to hit a home run or perform flawlessly in the field or exert perfect control on the pitcher's mound—then you could adjust your starting line-up so that it would be suitable for the opposing pitcher, the weather, and all the rest.

Ordinarily, it would take so long to cull out the necessary information that, instead, a manager would naturally prefer to use his own long experience, his own intuition and gut feelings, to direct his strategy, to tell him when to take out a pitcher or put him in, when to insert a pinch hitter, when to signal for a particular type of play, and so on. Naturally, some managers are better than others at these things, and some (McGraw, McCarthy, Mack, Stengel) are legendary. But we have the computer now, and, increasingly, computers are going to be used for these purposes. Statistics concerning current players can be fed into computer memories, and the proper programming will enable specific types of information to be filtered out in split seconds. The computer will then substitute for the manager's intuition in determining the line-up and strategy of the game.

Of course, when every team has its computer, a lot will depend at first on the efficiency of the program. In the long run, though, all the teams will have excellent programs, and what will happen then? It certainly won't mean that every game will end in a tie or that a team that is a shade better than the others in the league will win every game. The computer won't be predicting certainties, only probabilities. Hitter A usually does well against Pitcher B, but this time he gets 0 for 4. Maybe his shoulder is a little creaky; maybe he's got a tax problem that's gnawing at him; maybe he had a fight with his wife. Nor will a computer be able with certainty to predict injuries, lost tempers that lead to ejection from a game, and so on. All a computer can do is improve the odds of winning a game, and that's good enough.

Naturally, there may be attempts at trying to outwit the computer. Last-minute substitutions could render the opposition's starting line-up less effective. Of course, there could then be last-minute counter-substitutions, and it might be necessary to devise regulations that would carefully limit the ability of a manager to make substitutions during a game or within a certain period of time before a game.

Would sports writers set up their own computerized programs to determine in advance who was most likely to win a pennant or a World Series? Or for that matter, would professional gamblers and odds makers do the same? Would this application of computer technology to professional sports be counterproductive? Would the public become less interested in sports or in betting on the outcome if matters became more predictable? Or would there always be enough unpredictability to keep interest high? And would people derive particular excitement from beating the computer when low-ranking players on a particular team suddenly started playing over their heads through some unlooked-for inspirational factor? In any case, with the computer to help, the level of excellence in play is bound to improve, and that will be good for everyone.

31 DISASSEMBLING
THE ASSEMBLY LINE

BACK IN 1908, Henry Ford conceived the notion of an assembly line. Instead of having one man assemble a whole car—and another man simultaneously assemble a whole car—and a third man simultaneously assemble a whole car—each man performing many operations with many tools and many parts—Ford had all the men stand still while a car in the process of assembly moved along a track. Each man performed one operation with one set of tools on one part of the car. All the operations were conducted in sequence, so at the start of the line there was nothing but a bare frame, and at the end of the line, fully assembled cars drove off, one after the other after the other.

Efficiency was greatly increased. The assembly line was one demonstration of the kinds of technological advances that steadily raised the living standard of the American people and just as steadily raised the might and power of the American nation. But the assembly line had its bad points, as many were quick to point out. Reducing the job of a laborer to the endless repetition of one operation deprived him of an overall view of his work and of feelings of accomplishment and pride. It made of him an unimportant cog in a giant machine. Those of us who have seen Charlie Chaplin's film *Modern Times* know what I mean. We have watched Charlie tighten bolts all day long and be reduced to a cipher till he goes mad at last.

One would think there would be a loud cry of jubilation, consequently, if a way could be found to lift the load of assembly-line cipherism from the human being without affecting efficiency or the technological capacity of society. If this could be accomplished along with an actual increase in efficiency, that would surely be pure delight.

A way has been found, and it does increase efficiency. In principle, it is simple. If the assembly line reduces human beings to dull machines who perform mind-stultifying, repetitive tasks all day long, then invent an actual machine that will do the task instead. In practice, this is *not* simple. Even a repetitive task that places few demands upon the human brain turns out to be enormously complicated when we try to make a machine duplicate what a human being does. It is amazing how much in the way of human activity we take for granted until we try to design a machine to do it all.

If the machine were to perform Charlie's bolt-fastening job, it would have to see exactly where the bolts were, sense exactly how tight they were at every moment in order to know when to stop tightening, be able to notice a defective bolt and take action to replace it, adjust its own activity to the speeding and slowing of the assembly line for reasons not involving itself, and so on and so on.

But now we can do it all. The development of microcomputers has made it possible to equip a machine with enough of a "brain" to make it capable of fulfilling assembly-line require ments. An automobile can be built on the assembly line by a series of industrial robots, each designed to carry through a particular operation, each capable of sensing what needs to be sensed, each capable of adjusting its behavior through feed-back mechanisms.

Robots can work longer than human beings, do not suffer from boredom or fatigue, do not require coffee breaks or strike for higher pay or take dislikes to each other. And, on the whole, they make fewer mistakes and do a better job than human beings do. This is not just theory. In Japan, all-robot factories already exist and are turning out automobiles — the same automobiles that are driving American cars out of the market. The road of superior technology may lead to triumph for Japan where mere military force led to defeat forty years ago.

But what's to stop us from adopting more and better robots and from maintaining our technological superiority in the world? People! All the objections to the roboticization of humanity brought on by the assembly line vanish when it appears that human beings will be deprived of those roboticizing jobs. Better a human being made into a robot than a real robot depriving the human robot of a job.

We can understand. Being out of work, living on government

handouts, and feeling useless and unwanted is no great pleasure and can make even the assembly line seem like a paradise after a while. This means that when technological change comes, other things must change with it. There must be matching social change.

People have gone through the experience before. When the Industrial Revolution made Great Britain (for a while) the dominating world power, there was, at first, no matching social change. The early decades of the factory system saw indescribable misery inflicted on men and women and (far worse) on helpless children until reformers enforced changes.

We must learn from that epoch. As industry becomes roboticized, we must make that the occasion not of unemployment and welfare but of new education, new kinds of work, new sorts of creativity. If the assembly line roboticized human beings, then we must make its disassembly the occasion for the rehumanization of human beings.

32 TALKING TO MACHINES

ONE OF THE PROBLEMS of the modern computerized office is that it is difficult to persuade executives to use the machines as long as it is necessary to handle them manually. For one thing, it means learning a skill, and men high in their profession hate to run the risk of looking foolish by demonstrating a lack of skill or an inability to gain expertise. Then, too, running machines is, somehow, the task of underlings — assistants and secretaries. The busy executive does not do such things; he handles people. He does not dial the telephone; he says, "Get me William Smith on the phone!" He does not use a typewriter; he says, "Take a letter, Miss Jones."

Consequently, there are a number of firms who are doing their best to develop computers that will accept verbal commands. A computer that will respond to a hurried "Get me the latest figures on the Acme deal!" will be much more acceptable than one that has to be stroked, patted, and punched. It would also be better if in presenting a screen filled with figures or some printed sheets, the machine could accompany its presentation with a very polite, and even humble, "Yes, sir, Mr. Thwaitsfield. Here you are, sir."

It's not as easy as it sounds. Language is a complicated phenomenon. Even if we confine the whole matter to English and dismiss the difficulties of cross lingual communication, it remains complicated. Dialects and accents abound, not only on a regional basis, but on an individual basis. Every human being has some peculiarity of pronunciation.

I, for instance, was brought up in Brooklyn, and anyone can tell that in less than a minute. I can speak general English, enunciated with reasonable care, if I make a slight effort—but that effort must be made. If I relax, the words come out in peculiar sound combinations that are perfectly understandable to my beloved fellow Brooklynites but possibly to no one else. As an example, I might call out to some member of my family and ask the simple question, "Where are you?"—except that it comes out "Wheraya?" rhyming exactly with "Maria."

Then, too, even if every Anglophone (that is, every English-speaking individual) spoke perfect general English, perfectly enunciated, there are words and combinations of words that sound precisely alike but have widely different meanings, such as to—too—two, pray—prey, prince—prints, I scream—ice cream, a door—adore. There are also words that are not pronounced the same but are close enough so that they are easily slurred, such as divine—define, endure—injure, Bible—babble—bubble, and so on. It would take an enormous effort to devise a machine that could distinguish sounds well enough never to make a mistake among later, leader, latter, ladder, and liter.

I suspect that it would be comparatively easy to get a machine to understand 90 percent of what was spoken to it, yet almost impossible to get it to make sense out of the remaining 10 percent. The same might be true when we tried to get a computer to speak. I am sure that we will get rid of that annoying toneless metallic speech that computers always use in poor science fiction movies and teleplays, but even if we got reasonable pronunciation and intonation, could we be sure that the computer wouldn't stumble over sound-alike words? If it directs us to "the cinder track" and we hear this as "the center track," we are liable to be annoyed and to blame the machine rather than our own inattentiveness.

What is to be done? I suspect there will come a time when computer engineers will throw up their hands and declare that it will no longer pay to try to get the machines to come all the way over to the English language. People will have to compromise and go partway to the machine. In order to obtain the advantage of a conversational machine, one that hears us and replies without mistake, we might have to sacrifice some of the English language. If we want to keep "to," we will have to say "twice" instead of "two" and "also" instead of "too." We will have to say "mar" instead of "injure," or else say "survive" instead of "endure." We will have to work up a special pronunciation, perhaps, in which we eliminate the more uncomfortable consonant combinations, like the first "r" in "February."

In fact, we may have to develop a new language called Robotic, a form of basic English that will be sufficiently simplified in vocabulary, regularized in grammar, and eased in pronunciation for the rest of the world to want to learn it, too. We would then have a world language at last. But we'd better hurry, or Robotic will be a form of basic Japanese.

86

33 THE NEW PROFESSION

BACK IN 1940, I wrote a story in which the leading character was named Susan Calvin. (Good heavens, that's nearly half a century ago.) She was a "robopsychologist" by profession and knew everything there was to know about what made robots tick. It was a science fiction story, of course. I wrote other stories about Susan Calvin over the next few years, and as I described matters, she was born in 1982, went to Columbia, majored in robotics, and graduated in 2003. She went on to do graduate work and by 2010 was working at a firm called U.S. Robots and Mechanical Men, Inc. I didn't really take any of this seriously at the time I wrote it. What I was writing was "just science fiction."

Oddly enough, however, it's working out. Robots are in use on the assembly lines and are increasing in importance each year. The automobile companies are planning to install them in their factories by the tens of thousands before the end of this decade. Increasingly, they will appear elsewhere as well, while ever more complex and intelligent robots will be appearing on the drawing boards. Naturally, these robots are going to wipe out many jobs, but they are going to create jobs, too. The robots will have to be designed, in the first place. They will have to be constructed and installed. Then, since nothing is perfect, they will occasionally go wrong and have to be repaired. To keep the necessity for repair to a minimum, they will have to be intelligently maintained. They may even have to be modified to do their work differently on occasion.

To do all this, we will need a group of people whom we can call, in general, robot technicians. There are some estimates that by the time my fictional Susan Calvin gets out of college, there will be over 2 million robot technicians in the United States alone, and perhaps 6 million in the world generally. Susan won't be alone. To these technicians, suppose we add all the other people that will be employed by those rapidly growing industries that are directly or indirectly related to robotics. It may well turn out that the robots will create more jobs than they will wipe out —but, of course, the two sets of jobs will be different, which means there will be a difficult transition period in which those whose jobs have vanished are retrained so that they can fill new jobs that have appeared.

This may not be possible in every case, and there will have to be innovative social initiatives to take care of those who, because of age or temperament, cannot fit in to the rapidly changing economic scene.

In the past, advances in technology have always necessitated the upgrading of education. Agricultural laborers didn't have to be literate, but factory workers did, so once the Industrial Revolution came to pass, industrialized nations had to establish public schools for the mass education of their populations. There must now be a further advance in education to go along with the new high-tech economy. Education in science and technology will have to be taken more seriously and made lifelong, for advances will occur too rapidly for people to be able to rely solely on what they learned as youngsters.

Wait! I have mentioned robot technicians, but that is a general term. Susan Calvin was not a robot technician; she was, specifically, a *robopsychologist*. She dealt with robotic "intelligence," with robots' ways of "thinking." I have not yet heard anyone use that term in real life, but I think the time will come when it will be used, just as "robotics" was used after I had invented *that* term. After all, robot theoreticians are trying to develop robots that can see, that can understand verbal instructions, that can speak in reply. As robots are expected to do more and more tasks, more and more efficiently, and in a more and more versatile way, they will naturally seem more "intelligent." In fact, even now, there are scientists at MIT and elsewhere who are working very seriously on the question of "artificial intelligence."

Still, even if we design and construct robots that can do their jobs in such a way as to seem intelligent, it is scarcely likely that they will be intelligent in the same way that human beings are. For one thing, their "brains" will be constructed of materials different from the ones in our brains. For another, their brains will be made up of different components hooked together and organized in different ways, and will approach problems (very likely) in a totally different manner.

Robotic intelligence may be so different from human intelligence that it will take a new discipline—"robopsychology"— to deal with it. That is where Susan Calvin will come in. It is she and others like her who will deal with robots, where ordinary psychologists could not begin to do so. And this might turn out to be the most important aspect of robotics, for if we study in detail two entirely different kinds of intelligence, we may learn to understand intelligence in a much more general and fundamental way than is now possible. Specifically, we will learn more about *human* intelligence than may be possible to learn from human intelligence alone.

34 THE ROBOT AS ENEMY?

IT WAS BACK IN 1942 that I invented "the Three Laws of Robotics," and of these, the First Law is, of course, the most important. It goes as follows: "A robot may not injure a human being, or, through inaction, allow a human being to come to harm." In my stories, I always make it clear that the Laws, especially the First Law, are an inalienable part of all robots and that robots cannot and do not disobey them.

I also make it clear, though perhaps not as forcefully, that these Laws aren't *inherent* in robots. The ores and raw chemicals of which robots are formed do not already contain the Laws. The Laws are there only because they are deliberately added to the design of the robotic brain, that is, to the computers that control and direct robotic action. Robots can fail to possess the Laws, either because they are too simple and crude to be given behavior patterns sufficiently complex to obey them or because the people designing the robots deliberately choose not to include the Laws in their computerized makeup.

So far—and perhaps it will be so for a considerable time to come—it is the first of these alternatives that holds sway. Robots are simply too crude and primitive to be able to foresee that an act of theirs will harm a human being and to adjust their behavior to avoid that act. They are, so far, only computerized levers capable of a few types of rote behavior, and they are unable to step beyond the very narrow limits of their instructions. As a result, robots have already killed human beings, just as enormous numbers of noncomputerized machines have. It is deplorable but understandable, and we can suppose that as robots are developed with more elaborate sense perceptions and with the capability of more flexible responses, there will be an increasing likelihood of building safety factors into them that will be the equivalent of the Three Laws. (The Second Law makes the robot obedient within the limits of the First Law, and the Third makes the robot guard its own safety within the limits of the First and Second Laws.)

But what about the second alternative? Will human beings *deliberately* build robots without the Laws? I'm afraid that is a distinct possibility. People are already talking about Security

Robots. There could be robot guards patrolling the grounds of a building or even its hallways. The function of these robots could be to challenge any person entering the grounds or the building. Presumably, persons who belonged there, or who were invited there, would be carrying (or would be given) some card or other form of identification that would be recognized by the robot, who would then let them pass. In our security-conscious times, this might even seem a good thing. It would cut down on vandalism and terrorism and it would, after all, only be fulfilling the function of a trained guard dog.

But security breeds the desire for more security. Once a robot became capable of stopping an intruder, it might not be enough for it merely to sound an alarm. It would be tempting to endow the robot with the capability of ejecting the intruder, even if it would do injury in the process—just as a dog might injure you in going for your leg or throat. What would happen, though, when the chairman of the board found he had left his identifying card in his other pants and was too upset to leave the building fast enough to suit the robot? Or what if a child wandered into the building without the proper clearance? I suspect that if the robot roughed up the wrong person, there would be an immediate clamor to prevent a repetition of the error.

To go to a further extreme, there is talk of robot weapons: computerized planes, tanks, artillery, and so on, that would stalk the enemy relentlessly, with superhuman senses and stamina. It might be argued that this would be a way of sparing human beings. We could stay comfortably at home and let our intelligent machines do the fighting for us. If some of them were destroyed—well, they are only machines. This approach to warfare would be particularly useful if we had such machines and the enemy didn't.

But even so, could we be sure that our machines could always tell an enemy from a friend? Even when all our weapons are controlled by human hands and human brains, there is the problem of "friendly fire." American weapons can accidentally kill American soldiers or civilians and have actually done so in the past. This is human error, but nevertheless it's hard to take. But what if our *robot* weapons were to accidentally engage in "friendly fire" and wipe out American people, or even just American property? That would be far harder to take (especially if the enemy had worked out stratagems to confuse our robots and encourage them to hit our own side). No, I feel confident that attempts to use robots without safeguards won't work and that, in the end, we will come round to the Three Laws.

35 INTELLIGENCES TOGETHER

IN AN ESSAY ENTITLED "The Laws of Robotics" (see *Change!*, Houghton Mifflin Co., 1981), I mentioned the possibility that robots might become so intelligent that they would eventually replace us. I suggested, with a touch of cynicism, that in view of the human record, such a replacement might be a good thing. Since then, robots have rapidly become more and more important in industry, and, although they are as yet quite idiotic on the intelligence scale, they are advancing quickly.

Perhaps, then, we ought to take another look at the matter of robots (or computers—which are the actual driving mechanism of robots) replacing us. The outcome, of course, depends on how intelligent computers become and whether they will become so much more intelligent than we are that they will regard us as no more than pets, at best, or vermin, at worst.

This implies that intelligence is a simple thing that can be measured with something like a ruler or a thermometer (or an IQ test) and then expressed in a single number. If the average human being is measured as 100 on an overall intelligence scale, then as soon as the average computer passes 100, we will be in trouble.

Is that the way it works, though? Surely there must be considerable variety in such a subtle quality as intelligence; different species of it, so to speak. I presume it takes intelligence to write a coherent essay, to choose the right words, and to place them in the right order. I also presume it takes intelligence to study some intricate technical device, to see how it works and how it might be improved—or how it might be repaired if it had stopped working. As far as writing is concerned, my intelligence is extremely high; as far as tinkering is concerned, my intelligence is extremely low. Well, then, am I a genius or an imbecile? The answer is: neither. I'm just good at some things and not good at others—and that's true of every one of us.

Suppose, then, we think about the origins of both human intelligence and computer intelligence. The human brain is built up essentially of proteins and nucleic acids; it is the product of over 3 billion years of hit-or-miss evolution; and the driving forces of its development have been adaptation and survival. Computers, on the other hand, are built up essentially of metal and electron surges; they are the product of some forty years of deliberate human design and development; and the driving force of their development has been the human desire to meet perceived human needs. If there are many aspects and varieties of intelligence among human beings themselves, isn't it certain that human and computer intelligences are going to differ widely since they have originated and developed under such different circumstances, out of such different materials, and under the impulse of such different drives?

It would seem that computers, even comparatively simple and primitive specimens, are extraordinarily good in some ways. They possess capacious memories, have virtually instant and unfailing recall, and demonstrate the ability to carry through vast numbers of repetitive arithmetical operations without weariness or error. If that sort of thing is the measure of intelligence, then *already* computers are far more intelligent than we are. It is because they surpass us so greatly that we use them in a million different ways and know that our economy would fall apart if they all stopped working at once.

But such computer ability is not the *only* measure of intelligence. In fact, we consider that ability of so little value that no matter how quick a computer is and how impressive its solutions, we see it only as an overgrown slide rule with no true intelligence at all. What the human specialty seems to be, as far as intelligence is concerned, is the ability to see problems as a whole, to grasp solutions through intuition or insight; to see new combinations; to be able to make extraordinarily perceptive and creative guesses. Can't we program a computer to do the same thing? Not likely, for we don't know how *we* do it.

It would seem, then, that computers should get better and better in their variety of point-by-point, short-focus intelligence, and that human beings (thanks to increasing knowledge and understanding of the brain and the growing technology of genetic engineering) may improve in their own variety of whole-problem, long-focus intelligence. Each variety of intelligence has its advantages and, in combination, human intelligence and computer intelligence — each filling in the gaps and compensating for the weaknesses of the other — can advance far more rapidly than either one could alone. It will not be a case of competing and replacing at all, but of intelligences together, working more efficiently than either alone within the laws of nature.

36 LIGHTWEIGHT

FOR ABOUT THIRTY-FIVE HUNDRED YEARS, iron has been the most useful metal for general purposes. It is the cheapest of the metals, and if carbon is added to it in the proper proportions, it becomes steel and is unusually hard, tough, and strong. About the only serious disadvantage of iron and steel is that they rust, and the rust tends to flake away, exposing new surface to further rusting. There is enormous expense involved in trying to control the rusting, but even so, iron and steel remain the preferred metals for most construction.

Iron and steel are relatively dense metals. (In ordinary conversation, we would say they are "heavy.") A cubic foot of iron would weigh about 490 pounds, or about a quarter of a ton. This is not in itself a particular drawback, for the weight gives the iron construction a feeling of solidity and strength. However, there came a time when air travel was under development and it became necessary to combine strength with lightness. The earliest balloons were composed simply of fabric, but dirigibles had to take on a streamlined whalelike form and their balloons had to be encased in properly shaped rigid material—a metal, but not a dense metal, for every pound in the structure meant a pound less of payload.

People have known about aluminum ever since 1827. Aluminum is only about one-third as dense as iron. A cubic foot of aluminum weighs just under 170 pounds. What's more, although aluminum rusts, the rust forms a thin, transparent layer that clings to the aluminum surface and protects the metal underneath from further rusting. Finally, aluminum is the most common metal making up the Earth's crust, nearly twice as common as iron.

The chief problem, though, was the difficulty of getting aluminum metal out of its ores. It required so complex a procedure that aluminum was practically a precious metal. In 1856, its price was $90 per pound (in 1856 dollars). In 1886, however, an electrical method for obtaining pure aluminum in quantity was developed, and the price of the metal began to nose-dive till it was less than a dollar a pound.

Aluminum remained more expensive than iron, but its low density made it indispensable for the framework of dirigibles after such vehicles were first devised, in 1900. The first airplane flew in 1903. In order for the engine, and later the framework, to be light enough to enable flight, it had to be of aluminum, too. Aluminum itself turned out to be a little too soft to be useful for the purpose; too prone to be bent and twisted in hard service. In 1906, the Germans found that by the addition of a little copper and magnesium to the aluminum, an alloy could be formed that was much harder than pure aluminum and no denser. This was "duralumin."

The Germans used duralumin for the dirigibles that bombed London in World War II. When one of them crashed, the British analyzed the alloy and quickly developed duralumin of their own. Ever since, aluminum alloys have been indispensable for aircraft and for any other devices (from missiles to vacuum cleaners) in which a combination of lightness and strength is wanted, even if the alloys are accompanied by a bit more expense.

Is there, however, any way of further decreasing the density and weight of the metal framework of an aircraft (or other device) without decreasing its strength? There are, after all, some metals that are even less dense than aluminum. There is magnesium, for instance. Magnesium is only two-thirds as dense as aluminum, so that a cubic foot of magnesium would weigh only 110 pounds. Magnesium isn't strong enough to do the job by itself, but if some of it is added to aluminum, density goes down without decreasing strength much. Aluminum-magnesium alloys are about the best that can be supplied for aircraft now. The only metals less dense than magnesium are the alkali metals, which are too active chemically for use. However, the least active of these metals is also the least dense of *all* metals. It is lithium, which is only one-fifth as dense as aluminum, and, in fact, only half as dense as water. A cubic foot of it would weigh only 33 pounds.

Firms that produce aluminum are now experimenting with alloys containing lithium as well as magnesium in order to shave the weight a little further without loss of strength. The Soviet Union began using such alloys thirty years ago, but their alloys were not strong enough, and we're doing better now. It is very likely that aluminum-magnesium-lithium alloys are the best possible lightweights among metals. Such alloys can be handled with those techniques already commonly used in airplane manufacture.

Lying beyond these alloys are nonmetallic substances, such as ceramics, cements, plastics, and so on. They may prove to supply the necessary strength with less weight than would be involved in any possible metal or alloy. Such nonmetallic substances are so different from metals, however, that the technique of airplane manufacture would have to be altered in many basic ways. For a while, then, lithium may be the way to go.

---

37 GOLD!

SINCE CIVILIZATION BEGAN, gold has been an unbearably alluring object. It has been valued for its unparalleled beauty; for the ease with which it can be shaped into exquisite objects; for its resistance to rust, which enables it to last, unchanged and gleaming, indefinitely. Most of all, gold has been valued for its rarity. Its possession used to be such an unusual thing that it became a symbol of achievement and something to be lusted after. It made an ideal medium of exchange and became the embodiment of money and the epitome of wealth.

People would do anything, go anywhere, risk everything for the chance of obtaining gold. It was gold madness that drove on the early explorers of the New World, the pioneers who crossed the hostile Rockies in 1849 and who braved the frigid Klondike a half century later. And it is gold that sends people down 2.2 miles to the bottom of the deepest mine in the world. Yet with five thousand years of impassioned mining, all the gold that was ever dug out of the earth, if it could be assembled in one place, would make up a cube that was only 65 feet on each side. Let's put it another way. A football field (which most Americans are familiar with and can easily visualize) is a rectangle that is 360 feet long from goal post to goal post and 160 feet wide. If such a football field were covered with gold to a depth of 4¾ feet, that would represent all the gold that was ever mined.

Let's put it still another way. The five floors of the Pentagon have a floor area of 6,500,000 square feet. If the government, for some strange reason, decided to pave those floors with gold, all the gold ever mined would only be sufficient to produce a pavement that was half an inch thick. It wouldn't be a good idea, of course, for gold is very dense, and that half-inch layer of gold all over the Pentagon would weigh one hundred thousand tons and would put a serious strain on the building's structure. However, we may as well relax. Even the Pentagon wouldn't dream of trying it, for all that gold would cost (at today's prices) something like $1,300,000,000,000, or 1.3 trillion dollars.

At the present time, the world supply of gold is being increased by something like 1,420 tons each year, about half of it being produced by South Africa and an additional quarter by the

Soviet Union. At this rate, it will take only seventy more years to produce as much gold as was produced in the previous five thousand. Of course, production is not likely to hold up. People have been searching for gold in more and more inaccessible places and have been digging deeper and deeper. Very likely, the best sources have been emptied or are being emptied.

Yet why dig in the unyielding earth when the ocean is at our very doorstep? The ocean contains a little bit of every element —including even gold. Every ton of sea water contains about one five-millionth of an ounce of gold. Not much, we must admit, but there are an awful lot of tons of sea water in the ocean, so it contains about 10 million tons of gold altogether. This is one hundred times as much as all the gold ever mined and seven thousand times as much as we produce each year from the reluctant land. All we have to do is run sea water through some extracting mechanism and pull out the gold, little by little. All we have to do? It would take the complete extraction of gold from ninety-five tons of sea water to net us about one cent's worth of the metal, even at its present high price, and the handling of that much sea water would surely cost more than a penny even under the most favorable circumstances.

Still, what if we managed to produce, through genetic engineering, some form of alga or bacterium that could extract gold from sea water and store it within its cellular material? There are forms of seaweed that concentrate the rather rare iodine in this fashion, and other forms of sea life that concentrate the even rarer metal, vanadium. To be sure, vanadium in sea water is concentrated to five hundred times the extent that gold is, but it isn't beyond all conceptualization that cells having the capacity to filter out the occasional gold atom might be cultured. It would take time to accumulate a sizable amount of gold, but it might be less expensive to do it in this way than by purely chemical methods.

But then, if we succeeded, what would we do with all that

gold? It is one element that is scarcely used for any practical purpose. Almost all of it is squirreled away in vaults to serve as a psychological backing for financial transactions. Only about 20 percent of it is out in the open, mostly on display in jewelry and in the arts. And all this hypothetical success would be dependent on gold's rarity. If gold were considerably more common, it would be considerably less valued and considerably less useful as a psychological symbol for wealth. The mere announcement that a convenient way had been found of extracting gold from sea water might suffice to send the world economy into a tailspin.

So let's not do it.

38 THE SUPERHEAVIES

IN ALL THE UNIVERSE, as far as we know, there are only eighty-one stable elements—only eighty-one varieties of atom that, left to themselves, will remain as they are. Each has an "atomic number" that equals the number of protons present in the nucleus at the very center of the atom. The stable elements range from atomic number 1 to atomic number 83 (with atomic numbers 43 and 61 missing). In addition, there are two elements that are *almost* stable—thorium (atomic number 90) and uranium (atomic number 92). These do break down and change into elements of a smaller atomic number, but very slowly. It would take 4.5 billion years for half the uranium on Earth to break down, and 14 billion years for half the thorium to break down.

The last stable element to be discovered was rhenium (atomic number 75), in 1925. By that time, a few unstable elements had been discovered because they were produced when thorium and uranium broke down. They occurred in nature, but only in very small quantities. When isolated, they broke down quite rapidly. The most nearly stable variety of polonium (atomic number 84) was half broken down in only a hundred years. By 1937, all the elements from 1 to 92, stable and unstable, were known, except for 43, 61, 85, and 87. Then scientists began making new kinds of atoms in the laboratory. By 1945, those four missing elements had been made and studied, and were, of course, found to be unstable.

Meanwhile, scientists had also begun to put together elements having an atomic number greater than 92. This was done by firing small particles at large atomic nuclei. Under the proper conditions, the small particles would enter the nuclei and remain there, making them larger still. Once larger nuclei were formed, they would be bombarded with small particles and become still larger. In this way, as the years passed, elements with atomic numbers 93, 94, 95, and so on, all the way up to 104 and 105, were formed and studied.

The difficulty was that the higher the atomic number went, the more unstable the element was. Very few atoms were formed, and they didn't linger long. There was only a short time during which the bombardment could take place, and then even fewer atoms of still greater atomic number were formed, and these lingered even more briefly. It didn't look as though scientists would manage to go much beyond atomic number 105.

Yet they wanted to. Theoretical physicists had worked out the structure of the atomic nucleus and had discovered the rules that explained when elements would be stable. From the theories they had developed, it appeared that although elements grew more unstable as the atomic numbers grew higher and higher, there were exceptions. It seemed possible that elements of atomic numbers 110 and 114 might just possibly be stable—and that there might therefore be two new stable elements besides the 81 that had seemed to be the only stable ones in the universe. If the new elements weren't altogether stable, they might be almost stable, like thorium and uranium. These elements were called the "superheavies," and scientists were excited about their potential qualities. Isolating them was not only important in order to test the theory, but if they were stable, or nearly stable, they might eventually be prepared in considerable quantities, and who could tell what unusual properties and uses they might have?

But how could they be formed? It seemed to scientists that the only hope was to bombard atoms of high–atomic-number stable elements (which could be isolated in large quantities) with atoms of middle-sized elements. If the collision wasn't hard enough, the atoms would just bounce off each other. If it was too hard, then the large atom would break into pieces. If, however, the energy of the collision was just right, the two atoms would cling together and form a superheavy atom. The trick was to get the collision to be—just—hard—enough.

Toward the end of September 1982, German scientists bombarded bismuth atoms (atomic number 83) with iron atoms (atomic number 26) and seemed to have managed to get the two to stick together and form an element with an atomic number of 83 + 26, or 109. They formed a *single atom* of that element. It broke up almost at once just as theory had predicted such an atom *should* break up.

It's only one atom, but it indicates that the system works. If the proper atoms are chosen to smash together, and if they are made to collide with just the right energy, elements of atomic numbers 110 and 114 may be formed. If these are stable, or nearly stable, they will accumulate, and eventually we will have enough to test and study; and then, who knows? We may have two stable (or nearly stable) materials that, as far as we know, have not previously existed on Earth! And the only two that yet remain to be studied.

39 THE LOOKING-GLASS WORLD

THE CARBON ATOM is symmetrical. It can attach itself to four other atoms in such a way that any one of these attached atoms is equally distant from each of the other three. This is because the four carbon bonds are directed toward the vertices of a tetrahedron. If you're not familiar with tetrahedra, you can see this process, just the same, if you start with a marshmallow and insert three toothpicks to form a shallow three-legged stool. Then stick in a fourth toothpick pointing straight up. If you've done it correctly, you can turn the marshmallow any way and always have a three-legged stool with the fourth toothpick straight up. If you've not done it quite correctly, small adjustments will get you your symmetry.

You can imagine other atoms attached to the carbon atom by these bonds. If the four atoms are made up of no more than three different kinds (you can imagine maraschino cherries, pearl onions, and green olives, for instance), the symmetry holds. Attach two cherries, one onion, and one green olive to each of two marshmallows-with-toothpicks, and regardless of the order in which you add them, or to which toothpicks you add them, you will end up with the same model. If one model *looks* different from the other, you have but to turn one of them properly and it will then look just like the other. Suppose, though, that you have *four* different kinds of atoms attached: one cherry, one onion, one green olive, and one black olive. In that case, starting with a number of marshmallows-and-toothpicks, you will always end up with one or the other of two different arrangements, one of them being the mirror image of the other. There will be no way in which you can turn one arrangement into its mirror image, no matter how you twist and turn it, any more than you can turn a left-hand glove into a right-hand glove by twisting and turning it.

Any carbon atom with four different atoms (or atom groups) attached is an "asymmetric carbon atom" and exists in two mirror-image varieties. Chemists refer to these two varieties as D and L, but in popular language, they are sometimes called "right-handed" and "left-handed," respectively. For instance, all the amino acids in living tissue are of the left-handed configuration. They fit together to form left-handed protein molecules. Why left-handed? No reason! That's just the luck of the draw and the way life happened to evolve. All the amino acids could have been right-handed, and right-handed proteins might have formed from them—but they didn't. If we went through the looking glass, as Lewis Carroll's Alice did, we would be in a right-handed protein world, but, as it is, we are in our own left-handed protein world.

Some proteins are enzymes, and these are left-handed, too, of course. The enzymes control the chemical reactions of the body, making them move along speedily. In order for a chemical reaction to be speeded by an enzyme, the chemicals involved in the reaction must combine (very briefly) with the enzyme molecules. It is because the enzyme molecules and the chemicals they work on fit each other so snugly that enzymes perform their functions so well. Some of these chemicals, however, also exist in right-handed and left-handed forms. All sugars and starches, for instance, are right-handed, and individual compounds of this sort fit certain left-handed enzymes perfectly. For this reason, our right-handed sugars and starches are quickly digested, absorbed, and utilized.

But what if we could manufacture left-handed sugars or starches in the laboratory on a large scale, and cheaply? These would have the same chemical properties as the right-handed ones, except where other asymmetric molecules were involved. They would look like sugar and starch, behave like sugar and starch, taste like sugar and starch—but they would be part of the looking-glass world and would not fit our enzymes, so they would not be digested, absorbed, or utilized. They would pass through the alimentary canal untouched. Someday, then, we may use a left-handed sugar as a dietary additive. Provided that it reacted properly with the taste buds (which is not an absolutely sure bet), it would taste just like a normal sugar and would leave no unpleasant aftertaste, either. Nor should it have adverse side effects, though to be safe, this would have to be tested for. It should not even rot the teeth, since bacteria, with left-handed enzymes of their own, would find left-handed sugars useless as food and would leave them alone.

Because left-handed sugars (and starches, too) would not fit our enzymes, and therefore could not be used by the body, they would have no calories and could be the ultimate sweetening agent of the future. What's more, there might turn out to be uses for other compounds of the looking-glass world; compounds that would seem, taste, smell, and feel perfectly normal to us but would be utterly exotic and untouchable to our enzymes.

40 FOUR TIMES FOUR TIMES FOUR—

DEOXYRIBONUCLEIC ACID (DNA) is the master molecule of life. Its structure controls the physical characteristics of every living thing. Each living thing is at least slightly different from every other living thing, and in some cases *very* different (a son is slightly different in appearance from his father and *very* different in appearance from an oak tree), and this must come about because the DNA molecule is different in structure in different living things. To see how that can be, we must understand that the DNA molecule is very large and is built up of a chain of comparatively small molecules called nucleotides. There are four different nucleotides, which can be referred to by the initials of their chemical names as A, C, G, and T.

If you start with one nucleotide, it can be any one of the four, but suppose you want to make a chain of two-nucleotide combinations. You can use any of the four, and to each one add any one of the four. You end up with four times four, or sixteen, two-nucleotide combinations. They are AA, AC, AG, AT, CA, CC, CG, CT, GA, GC, GG, GT, TA, TC, TG, and TT. If you want a three-nucleotide combination, you can take each of the sixteen different two-nucleotide combinations and add any of the four nucleotides to it. That means a total of four times four times four, or sixty-four different three-nucleotide combinations. I won't list them, but you can if you wish to. It is tedious to do so, but not difficult.

In short, if you build up a nucleotide chain of any number, you must take that number of fours and multiply them together. That will give you the total number of different nucleotide arrangements you can have. The number grows very large in no time at all, larger than you would think. In fact, it grows larger than anyone can imagine. For instance, the DNA molecules in a small virus (the smallest living thing) can be made up of a chain of about 1,500 nucleotides. To get the total number of arrangements, you can write down 1,500 fours and multiply them together: four times four times four times four times four—and so on. If you multiply those 1,500 fours (and don't worry, you won't) you'll find the answer to be 10^{900}; that is, a 1 followed by 900 zeros; 100000000000—nine hundred of them. What does that mean? Well, a billion is 10^9—a 1 followed by 9 zeros. A billion billion is 10^{18}—a 1 followed by 18 zeros. A billion

billion billion is 10^{27}—a 1 followed by 27 zeros. Therefore, 10^{900} is a billion billion billion billion billion—well, you would have to write a hundred "billions" in a row, and I don't want to do that.

Let's try something else. Suppose every single particle in the universe were a DNA molecule; every proton, every neutron, every electron, every neutrino, every photon in every planet and star and dust cloud in every galaxy in the universe. And let's suppose that all these DNA molecules were different. Would there be enough different DNA molecules to go around? The total number of particles in the universe is about 10^{100}, but that is so much smaller than 10^{900} that it is practically zero. If every particle in the universe were a different DNA molecule, the number of molecules that would be used would be a billionth of a billionth of a billionth of a—well, eighty-eight "billionths of a" would have to be included in the series to arrive at a fraction of the total number.

Suppose, though, that every single second all those DNA molecules changed into other, hitherto unused DNA molecules, and that this had been going on ever since the stars and galaxies first formed and life became possible, and would continue going on till the stars and galaxies were all dead and life were no longer possible. The lifetime of the stars, in general (considering that new ones are being formed all the time), is thought to be something like ten thousand billion billion years. In all that time, 10^{130} different DNA molecules would have appeared—a 1 followed by 130 zeros. Yet that is still practically zero compared to the total number. It is only a billionth of a billionth of a billionth—(eighty-five "billionths of a" in a row) of the total number. And remember, we are speaking of a very small DNA molecule. The molecules in the cells of organisms such as ourselves are very much longer than those in a small virus. The number of different varieties of that much longer molecule is vastly larger than the one we have been talking about.

No wonder, then, that every human being looks and sounds different from every other one, and that individuals of every species look different from one another. No wonder that evolution has developed tens of millions of different species (most now extinct) in the 3.5 billion years that life has existed. And how silly it is, thus, to suppose that intelligent life on other worlds will look just like human beings, except for being colored green or having antennae or bulging foreheads. With so many different DNA molecules that might turn up as a result of random changes ("mutations"), the one thing we can be sure of is that any extraterrestrial form of life will be totally different from anything here on Earth.

What a pity we haven't come across any extraterrestrial life yet. How fascinated biologists would be if we did.

41 THE DRYING PUDDLE

THE JORDAN RIVER, which forms the eastern boundary of the nation of Israel, is one of the most famous short rivers in the world because of its Biblical associations. It runs through a deep canyon, the southern portion of which is well below sea level, and it never reaches the ocean. Instead, the river water collects and forms a lake at the lowest portion of the canyon, a region surrounded on all sides by high ground that prevents the water from flowing out. The lake does not rise till it overflows the high ground, because it is dry and hot in the region, so evaporation is rapid and balances the water that enters from the Jordan.

The Jordan constantly adds water to the lake, together with small traces of salt that the river dissolves out of its banks. The water evaporates from the surface of the lake, but the salt does not. The small traces of salt have accumulated steadily over uncounted thousands of years, until the lake is now about 25 percent salt. It is the saltiest natural body of water of any size in the world. Because it is salty, the lake is called a sea, and because there is so *much* salt in it, nothing can live in its waters and it is called the Dead Sea.

Like the Jordan River, the Dead Sea is famous because of its Biblical associations. The cities of Sodom and Gomorrah, which according to the Bible were destroyed in a rain of fire for their sins, are supposed to have existed near the southern end of the Dead Sea. Despite its fame, however, the Dead Sea is quite small. In 1930, its area was only about 370 square miles, which is less than the area of Los Angeles. It was only 50 miles long and 11 miles wide at its broadest. Its depth, however, was over 1,000 feet on the average, and it contained a total of 75 cubic miles of water. Its shore was 1,283 feet *below* sea level. There is no dry land anywhere in the world that is lower than the shore of the Dead Sea.

As it happens, the Dead Sea is a puddle of water that is slowly drying up. The trouble is that Israel is using the water of the Jordan River for irrigation purposes, so less of that water ends up in the Dead Sea. Jordan uses the Yarmuk River (which flows into the Jordan) for the same purpose. When the irrigation projects are all completed, the Jordan will carry only one-sixth as much water into the Dead Sea as it did in 1930. In the past half century, then, the area of the Dead Sea has shrunk by 20 percent. It is now only 309 square miles in area, just about the size of the five boroughs of New York City, largely because the water level has dropped by 36 feet, so that the shallow southern end has completely dried out. This isn't quite as bad as it sounds, since only 2 cubic miles of the Dead Sea's 75 have evaporated. Nevertheless, the process is continuing; the puddle continues to dry up.

The Dead Sea is only about 60 miles or so from the Mediterranean Sea, and the Israeli government is thinking of setting up a pumping station at the Mediterranean near the southern end of its territory. It will pump water to a height of 300 feet, and that water will then be allowed to flow eastward and downward, to its original level, along a canal, and then through a long tunnel under the hills. The water will finally emerge on the hillside overlooking the southwestern tip of the present Dead Sea. Such a scheme, if it is carried through, will cost perhaps 1.3 billion dollars. Why bother?

Well, the water will emerge from the hillside, 1,300 feet above the surface of the Dead Sea, and will be a natural waterfall. The falling water would spin turbines and generate electricity. The Israelis calculate that they can generate 850,000 kilowatts of energy in this way—which would be very useful for a nation forced to depend on oil imports that are not easy to get under present circumstances. The plan is to allow about 55 billion cubic feet of Mediterranean water to pour into the Dead Sea each year. This will be enough to allow the Dead Sea to fill up slowly over the years. In about twenty years, it will have filled up to the level at which it stood in 1930, and the region to the south, newly exposed, will be water-covered again.

Once the Dead Sea is back to its 1930 level, the influx of water will be cut to about 42 billion cubic feet per year, which, together with the water entering the Dead Sea from the Jordan River, would just balance evaporation, and the puddle would then neither dry nor flood. It is an ambitious project, and, considering the situation in the Middle East, there will probably be political and diplomatic complications that will be more difficult to solve than the engineering problems will be.

NOTE: *This essay appeared in September 1983, and my final foreboding was, of course, correct. Until the Middle East begins to experience the novelty of sanity, I doubt that anything constructive can be done in the region.*

42 BACKWARD! TURN BACKWARD!

INCREASINGLY, AS THE YEARS PASS, the world is going to worry about food. Population is still increasing at a rate of a couple of hundred thousand mouths each day, and we are just one rotten harvest away from world famine. So far, when one region does poorly, another region (usually the United States) can take up the slack, but if North America should suffer a drastic shortfall just once, the whole world would be in trouble. The governments of the world know this, and there is everywhere deep concern about increasing the level of food production. One way out of the problem is finding new land to put under the plough. The land is there, to be sure, and if it hasn't been used before now, there is a very good reason: often, it lacks rainfall.

The Soviet Union knows this well. In Tsarist days, Russia suffered periodic famines. The population is now some 70 million greater than at the time of the revolution, and famines don't take place. One reason is that the Soviet Union has been growing crops in central Asia to add to the yield of the old breadbaskets in the Ukraine and elsewhere. The difficulty is that the new lands are semidesert, so the Soviet Union lives always on the edge of poor harvests. More often than not, it has to import grain. The Soviet Union doesn't enjoy this any more than we would under similar circumstances. The only thing it can do about it is to irrigate the central Asian lands as much as possible. This means farmers must use water from the Volga River, which flows into the Caspian Sea, and from two central Asian rivers, the Syr Darya and Amu Darya, which flow into the Aral Sea. The farmers are using so much river water that the Caspian Sea and the Aral Sea are shrinking, and yet more water still is needed. What makes the situation particularly exasperating is that there are large rivers, the Pechora, the Ob, the Yenisey, which flow through frozen tundra into the Arctic Ocean. The lands they drain cannot be used for agriculture, and the water simply goes to waste (as far as human beings are concerned, anyway).

Is there any way of making these rivers flow backward— southward instead of northward? Could the Pechora turn its waters partly into the Volga and down into the Caspian? Could the Ob and Yenisey pour at least some of their flow into the Aral Sea? If so, might not the steppes of Kazakhstan be made into an ocean of grain, rivaling and even surpassing the North American prairies in extent, fertility, and yield? The Soviet Union has been weighing the possibilities of resculpturing the face of the Earth for years, but there are two huge questions. One is just how one goes about making rivers go where *you* want them to go instead of where the terrain sends them. The other is what would happen if you should chance to succeed.

Clearly, to turn the rivers around, you would have to construct dams to prevent the flow in the natural direction (at least in part), and you would have to dig canals in order to create a new path for the river to use. You would also perhaps need pumps to drive the water uphill a bit. (If the direction you wanted wasn't uphill to begin with, the river would take that path naturally.) In years past, the Soviet Union talked of using nuclear explosions to produce the huge displacements of earth that would be required, but nowadays the very thought of such explosions in open air is taboo. Undoubtedly, however, nonnuclear techniques would suffice if enough muscle and enough time were applied to the matter.

Next, suppose that the effort were made and that it succeeded. Every gallon of water that flowed southward to irrigate the croplands of the steppes would then be a gallon of water that didn't reach the Arctic Ocean. Would that matter? We can't tell for sure whether or not it would, but there are reasons to think it might. The Soviet rivers that flow into the Arctic pour fresh water into the sea. The fresh water is less dense than the salt water already in the Arctic and floats on it. It acts as a ceiling that prevents warmer water flowing in from the south from moving upward to melt the year-round ice cover of the Arctic Ocean.

If the freshwater flow were decreased, the warm water flow might be more effective and the Arctic ice might dwindle. This could mean that more water vapor would appear in the air above the Arctic regions and that there would be more precipitation. This in turn would mean a thick layer of winter snow in Scandinavia and Canada, as well as Siberia, a layer that would not entirely melt in the summer. The glaciers would form and expand; a new Ice Age would be on the way.

Grandiose schemes, in short, could have grandiose effects that would by no means necessarily be confined to the region where the schemes were put into effect. This means that the United States and Western Europe should be as interested in this scheme as ever the Soviet Union might be. Here is an example of how important the global view is. What goes on within a particular nation is by no means necessarily the concern of that nation alone.

43 HEY, ALASKA, HERE WE COME!

EVERY ONCE IN A WHILE, when there are rumors of an upcoming earthquake or comet or planetary line-up or almost anything, there are predictions that California will fall into the sea and various Californian innocents will make for the hills.

This is impossible, of course. California cannot fall into the sea.

Most of California, along with all the rest of North America (plus Greenland and the western half of the Atlantic Ocean), is affixed to a large hunk of the Earth's crust that is called "the North American Plate." The coastal regions of southern California up to a point just south of San Francisco are not part of the North American Plate, however. They are part of "the Pacific Plate," which includes most of the basin of the Pacific Ocean. These two plates, together with four or five other large plates plus half a dozen smaller plates here and there, make up the entire crust of the Earth, and all fit together snugly. There isn't room for much movement, but the plates manage to shift about very slowly; they move apart in some places, come together in others, or slide along sideways in still others.

For instance, Antarctica once was adjoined to southern Africa, many millions of years ago, and experienced salubrious weather. (There are fossils of ancient amphibians in Antarctic soil, and these amphibians, when alive, couldn't possibly have endured anything but mild weather; and there are Antarctic coal deposits, too.) The "Antarctica Plate" drifted slowly south, however, carrying Antarctica over the South Pole where it lies today, buried under several miles of ice. Then, too, India was adjacent to Africa once and was carried northward very slowly, till it collided with southern Asia, to which it now clings. As it collided, the south Asian coast wrinkled under the pressure, producing the biggest wrinkles on Earth's land surface, in fact, in the form of the Himalayan mountain complex that bounds India on the north.

The North American Plate and the Pacific Plate are neither pulling apart nor colliding, at least not at the moment. Instead, the Pacific Plate is slowly rotating so that its California edge is sliding northward along its junction line with the relatively motionless North American Plate. The part of the junction line which marches along the southern California coastline is the well-known San Andreas Fault. What it amounts to is that the southern California coastal region is moving northward past the rest of California (and North America). If the junction line were absolutely smooth and slippery, the movement would be gentle and would go unfelt, but, of course, the junction line is nothing of the sort. There is enormous friction and the movement snags, therefore, until the accumulating pressure of the overall turning of the Pacific Plate forces it free in the sudden jerk we call an earthquake. An earthquake can do a lot of damage, but what it *can't* do is send the California coast hurtling into the Pacific Ocean. The coast moves *with* the Pacific Ocean basin. The whole thing is turning as a unit.

The movement along the San Andreas Fault, while slow, is not totally insignificant. It amounts to about 10 centimeters (4 inches) a year. In other words, Los Angeles is 20.6 meters (23 yards) closer to San Francisco now than it was at the time the Declaration of Independence was signed. That's still not much, to be sure, but that movement was enough to cause some very serious earthquakes, including the most famous in American history, the 1906 San Francisco quake, and it will inevitably produce earthquakes in the future, too.

If this movement continues at its present rate, then in about 10 million years, Los Angeles will be located just west of San Francisco and the two can then form a single municipal government. That won't last forever, of course, for Los Angeles will continue moving northward (assuming that the Pacific Plate continues to turn). Eventually, 40 million years from now, the southern California coast will be sliding along the southern coast of Alaska, and it may possibly stick there, as India is sticking to Asia.

To dyed-in-the-wool Californians, this may be a bitter fate indeed. It may seem to them far preferable that California fall into the sea with a mighty splash than that it end up as an ignominious ice box, with all the beautiful people sitting on the beach in their fur coats, watching the icebergs drift by. Still, I *did* say that this would take place 40 million years from now. Chances are the human species won't last that long, so Californians will be spared that Alaskan humiliation.

44 THE CHANGING DAY

PRESIDENTIAL ELECTIONS COME in Leap Year. You'll admit 1980 was a Leap Year, and 1976, 1972, and 1968. Presidential elections come every four years and so do Leap Years, and that's that.

Yet in 1800, Thomas Jefferson was elected, and in 1900, William McKinley was elected, and neither year was a Leap Year. Leap Years do come every four years, but in the course of each four-hundred-year period, there are three times when we miss a Leap Year and have to wait *eight years* for one. The years that end in oo and that aren't divisible by four hundred are *not* Leap Years. The year 2000 will be a Leap Year, but 2100, 2200, and 2300 will *not*, any more than 1900 or 1800 was.

Why is this? Well, if the year were exactly 365.25 days long, then we would add a day to the 365 every four years to keep the calendar exactly even with the Sun. We would have one Leap Year every four years, or one hundred Leap Years every four hundred years. However, the actual length of the year is 365.2422 days, which is just about 365 and 97/400 days, so we need ninety-seven Leap Years every four hundred years, not one hundred of them. We have to skip three of the one-every-four in each four hundred-year interval, which makes for an additional complication to the already complicated calendar. Is there any way of stopping this complication?

116

One thing we can do is simply wait. The day does not stay the same length. The Moon sets up tides on the Earth. There is a bulge in the water on opposite sides of the Earth, and as our planet turns, the various land surfaces move through each bulge. The water works its way up every shore, then down, twice a day, and each time this happens there is friction of water against land. This acts as a brake on the Earth's rotation and gradually slows it down. The Moon's tidal influence also causes the solid rock of Earth to bulge a few inches upward on opposite sides of the Earth. The rock pushes up, then down, twice each day, and as layers of rock rub and slide against other layers of rock, that also creates friction and slows the Earth's rotation.

The Earth rotated more quickly in the past. There was a time when the day was only 23 hours, 59 minutes, and 58 seconds long. With the day 2 seconds shorter than it is now, there were exactly 365.25 of those slightly shorter days in the year, and we could have a Leap Year every four years without any interruption. The Earth will rotate more slowly in the future. There will come a time when the day will be 24 hours and 10 seconds long. With those extra 10 seconds per day, the year will be exactly 365.2 days long, and we will need a Leap Year every five years without exception. There will even come a time when the day will be 24 hours and 57.3 seconds long, and then there will be exactly 365 days in the year and we won't need *any* Leap Years at all (at least until the day lengthens still more and we have to *subtract* a day from the 365 every once in a while). *When* in the past, though, and *when* in the future are we talking about?

As it happens, the Earth's turning has a great deal of energy to it, and the brake applied by tidal action is very weak in comparison. At the present rate at which the tidal action is braking the Earth, it succeeds in lengthening the day by only 1 second after 62,500 years. To put it another way, each day is a rather unnoticeable 1/23,000,000 of a second longer than the day before. (There are greater changes than that from day to day for other reasons, but these other changes swing back and forth, while the 1/23,000,000 of a second per day is a steady change toward lengthening days.) This means that it was about 125,000 years ago that we had Leap Year every four years without exception, when Neanderthals were the highest form of humanity. And it won't be till 625,000 years in the future that we'll need Leap Years every five years without exception, and about 3,580,000 years in the future before we will be able to do away with Leap Years altogether—for a while.

So the present system doesn't require a rapid adjustment. As the day grows longer, we will, every once in a long while, have to drop a fourth Leap Year in the four-hundred-year interval. With time, it will have to be done at decreasing intervals. Finally, in about thirty-eight thousand years, we'll have to drop *four* Leap Years in *every* four-hundred-year interval, or one every century. In other words, *every* oo year and not just three out of four will have to be *not* a Leap Year. There's no need to worry about it, though. Astronomers will keep track, you may be sure —if civilization lasts long enough.

---

45 ON THE RISE

IT'S LUCKY THERE'S a little carbon dioxide in the atmosphere. It is of no direct use to us, but plants live on it and use it as the raw material out of which they build their tissues. Animals (including us) could not live without this activity of plants. For every million pounds of air in the world, there are 340 pounds of carbon dioxide. This means that carbon dioxide makes up 0.034 percent of the atmosphere, or 340 parts per million. That's not much, but it's enough to keep the plant world going.

Plants consume carbon dioxide and give off oxygen, whereas animals consume oxygen and give off carbon dioxide. Ideally, this means the amount of oxygen and carbon dioxide in the atmosphere should remain steady. That's not so, however. Carbon dioxide is on the rise. Every year it's a little higher than the year before.

There's no mystery as to why this should be. Human beings are burning coal and oil constantly, and when these are burned, carbon dioxide is formed and is discharged into the atmosphere. Right now, about 5 billion tons of carbon dioxide are discharged into the atmosphere every year as a result of the burning of fuel the world over. (One quarter of this total is contributed by the United States.) About half of this is dissolved in the ocean or reacts with the rocks, but the other half remains in the atmosphere.

We don't really know for certain what the atmospheric content of carbon dioxide was before large quantities of fuel began to be burned, because reliable measurements weren't made before 1958. We can estimate, though, the total amount of fuel that has been burned, and if the carbon dioxide it produced is subtracted from the total, it would seem that the measurement might have been 290 parts per million before the Industrial Revolution started two hundred years ago.

There are some more direct ways of estimating what past air was like. Deep in the ice caps that cover Greenland and Antarctica is ice that froze a thousand years ago or so and has been untouched ever since. Trapped in that ice are tiny gas bubbles, and these contain bits of the atmosphere as it existed at that time. From the analysis of this trapped air and also from a consideration of tree-ring data, it would seem that the preindustrial air had only 270 parts per million of carbon dioxide. If so, a rise of 20 parts per million in the last couple of centuries was not the result of the burning of fuel. Instead, it seems likely that this part of the rise was contributed by the cutting down of forests and their replacement by farms and pastures. Forests are great consumers of carbon dioxide. The crops and grass that replace them also consume carbon dioxide but not to as great an extent. Consequently, the destruction of forests results in a rise in atmospheric carbon dioxide.

Actually, if we stop to think about it, this isn't much. All the burning and all the forest cutting we have done in the industrial era has been damped by the hugeness of Earth's atmosphere. It has resulted in only a 25 percent rise in the tiny quantity of carbon dioxide present in the air. However, we're not stopping. In fact, we're burning fuel in greater and greater amounts each year, and our destruction of forests is accelerating, too. By some estimates, the quantity of carbon dioxide in the atmosphere will double over the next century and will be 680 parts per million in 2085.

This is still not much. We won't have trouble breathing, and plants will find the air richer and, all other things being equal, will grow faster. However, carbon dioxide has the ability to hold heat even in small quantities. If its quantity in the atmosphere doubles, it may raise Earth's average temperature just a bit. Again, an additional degree or two may not affect us directly (especially if we have air conditioners), but it would have an effect on the climate, and probably not a good one. A bit more heat might shift the rain bands poleward and hasten the way in which deserts are spreading in those latitudes where they are common. The polar ice caps would melt to at least some extent, and the sea level would rise, to the discomfort of those who reside in coastal areas.

As yet, we don't know for certain whether this will happen, since we can't predict exactly to what extent the ocean will dissolve additional carbon dioxide, to what extent exposed rocks will react with it, and to what extent faster growing plants will consume it. Still, there are grounds for uneasiness. If the time should come when we decided that the carbon dioxide level in the atmosphere was beginning to produce uncomfortable effects, it might be too late to do anything about it. We can easily raise the level of carbon dioxide, but we don't know of any easy way to lower it. It's easy to change forests into farmland, but very difficult to change needed farmland back into forests. And it's easy to burn coal and oil, but just about impossible to unburn them.

So we must decide as soon as possible how long we can allow the amount of carbon dioxide to be on the rise, and then act upon that.

46 THE ONE-TWO PUNCH

EVER SINCE MAY 18, 1980, when Mount Saint Helens blew its top, Americans have been interested in volcanoes. They are no longer a phenomenon that need be thought of as existing only in distant lands. Studying volcanoes isn't easy, however. While they are in progress, they are not safe to approach; and before and after the actual eruption, we aren't very sure as to *what* is going on underneath the ground, where the action is. All this has increased interest, too, in past volcanic eruptions, some of which were so huge that, compared to them, Mount Saint Helens was just a firecracker.

The most spectacular volcanic eruption in historic times, as an example, took place in the Aegean Sea, midway between Greece and Turkey. The volcano in question jutted up above sea level and made up the island we now call Thera. That island was once round, with a peak at the center, and it was, in ancient times, the site of a flourishing city that had close cultural connections with the advanced Minoan civilization on the large island of Crete, sixty-five miles south of Thera. After archeologists uncovered the ruins of the old Cretan civilization, it became clear that the Minoan way of life was advanced and sophisticated. Crete was the first nation to have a strong navy that protected it effectively from all enemies, so its cities needed no walls and its citizens had to fight no wars. About 1500 B.C., however, it all came to an end suddenly. The cities were shaken down, and the palaces were gutted by fire. When things settled down again, the Minoan civilization had ended and Crete was ruled by the less civilized Mycenaeans from the Greek mainland.

There was no way of telling at first just what had brought this about. The logical assumption was that the Mycenaeans (who, three centuries later, were to fight the famous Trojan war) had simply assaulted and conquered Crete; but Crete should have been too powerful for these semibarbarians. Beginning in 1966, though, archeologists digging up sites on Thera found evidence of an enormous volcanic explosion's having occurred at just about 1500 B.C., an explosion that had blown the entire center of the island into the stratosphere, so that what is left of it is shaped like pieces of a doughnut with the sea in the center. The volcanic eruption must have set off a tsunami (or tidal wave) which washed up against surrounding shores catastrophically. A rain of ashes and vapor must have inundated surrounding lands, and Crete must have received the worst of the blow. With the island laid low, its proud ships destroyed, its unarmed cities flattened, and much of its population dead, it easily fell prey to the Mycenaeans. Then, too, the proud city of Thera, destroyed in a day and sunk beneath the sea, must have given rise to the legend of Atlantis, which was immortalized over eleven centuries later by Plato.

But could a single blow have toppled the proud Minoan civilization? For the answer, archeologists have now turned to the new technique of "archeomagnetism." When volcanic lava is molten, it is too hot to show any magnetic properties. As it cools and solidifies, though, some of its molecules line up parallel to the Earth's magnetic field. If there were a single volcanic eruption and all the lava cooled off pretty much at the same time (over a period of days or even months), then all the rocks would have their magnetism lined up in the same way. However, the direction of the Earth's magnetic field, and its intensity, too, change slowly with time. If rocks in one place show magnetic fields that line up distinctly differently from those in another place, then two eruptions must have occurred years apart, and it is possible to calculate by how many years they were separated.

From the sites of the ruined cities of Minoan Crete, samples of solidified lava have been taken, and in recent years it was found that the lowest and, therefore, oldest lava samples from central Crete have magnetic properties quite different from those in eastern Crete. It now seems very likely that Crete suffered from the results of not one but two explosions of Thera. The first was severe enough, a volcanic eruption that showered central Crete with destruction and set off earthquakes that destroyed cities by shock and by fire. Then, anywhere from ten to thirty years later, came the second and much larger eruption that tore Thera apart. Crete got a worse blow than before and this time, eastern Crete, which had escaped the first disaster, was solidly hit. Under the force of this deadly one-two punch, Crete lay prostrate and helpless. Greece itself suffered much less, for it was upwind from Thera, and Mycenaean raiders invaded and took over Crete.

The lava flows of other past eruptions may be studied in this way, too, and it is to be hoped that they will add significantly to our knowledge of how volcanoes behave. This may strengthen our ability to predict eruptions and to evacuate potential victims in time to save many lives.

47 THE MISSING CRATER

IN MY ESSAY "Nearly Wiped Out" (see *Change!*, Houghton Mifflin Co., 1981), I described the finding of a layer of sediment deep underground that was surprisingly high in the rare element iridium. The sediment was about 65 million years old and dated back to the time of the extinction of the dinosaurs. Scientists at once suspected that the unusual nature of the sediment had something to do with the extinction. The Sun is richer in iridium than the Earth's crust is, so one suspicion was that the Sun had undergone a tiny explosion 70 million years ago, enough of one to shower the Earth briefly with solar matter (with its high levels of iridium) and to produce a blast of heat that killed off not only the dinosaurs but also three-quarters of all the species of life that then existed on Earth.

In the year and a half since that essay appeared, scientific opinion has veered. The layer of iridium-high sediment has been found in many places on Earth, and it is always the same age. Other metallic elements have also been found to be high in those sediments, and the various metals that are abnormally high in the sediments are found to exist in the same proportions as they exist in meteorites. Now the feeling is that a gigantic meteorite struck Earth 70 million years ago.

It was actually an asteroid, probably; one that was 6 miles or so in diameter. Such an asteroid would have gouged out a crater 110 miles across, covering an area equal in size to the state of New Hampshire. The heat of impact would have disintegrated and vaporized the asteroid and sent it up into the atmosphere in a cloud of fine dust that eventually settled out and produced the layer of iridium-high material all over the Earth. Along with the asteroid dust, vast amounts of debris from the Earth's crust were blasted upward. Altogether enough dust was hurled into the stratosphere to spread out and circle the entire globe thickly enough to block sunlight from the Earth's surface. According to some estimates, the light that reached the Earth from the Sun was only one-tenth as bright as the full Moon. And that darkness may have continued for *three years* before enough dust had settled out to bring about the restoration of a sizable fraction of the Sun's light. Remember, too, that the absence of sunlight meant the bitter deep freeze of a three-year-long winter.

With three years of frozen darkness, almost the whole plant world died, then most of the animals that live on plants, then most of the animals that live on animals. All the dinosaurs were killed off, as were many other species of animals both large and small. Some plants survived the three years of darkness in the form of spores and seeds and root systems. Some animals survived by living on frozen carcasses and on plant remnants. When the dust finally thinned and the darkness lifted, the Earth slowly grew green with plants again, and animals multiplied—but only the survivors in both cases. The species that had vanished never returned. But, in that case, where is the crater? There is no sign of a huge crater, one that is as large as New Hampshire, anywhere on Earth. Of course, it wouldn't exist in its original form after millions of years of erosion by wind, water, and the action of life forms, but it would leave some sort of rounded formation outlined by broken rock strata, and it might well be filled with water to form a circular lake. Older and smaller craters have been detected, but not this one.

We must remember, though, that 70 percent of the Earth's surface is water-covered, so there is a 7-to-3 chance that the asteroid hit the ocean, streaked through the water (setting up a huge tidal wave that blasted the nearby coasts with the biggest splash of all time, perhaps), and gouged out the sea bottom. If so, the huge crater was formed on the sea floor. It would certainly have filled up with sediment since then, but signs of its presence may still exist.

We have bathyscaphes now that are capable of exploring the deepest parts of the ocean floor. Might it not be advisable to build more and better bathyscaphes with which to explore the sea bottom in detail? To be sure, we have learned much about the sea bottom in a general way in the past thirty years. We know of its vast mountain chains and enormous canyons. Still, we can't possibly learn the fine details unless we go down there and observe at close range, either with our instruments or with our eyes.

We would surely find new forms of life. We would surely learn more about the metallic nodules on the sea floor, nodules that may yet become an important resource for our industries (see essay 20). And we might find a sign of a large crater somewhere that would offer the final bit of proof of a colossal catastrophe that came within a hair's breadth of sterilizing the Earth.

And if so, that would lend force to my suggestion that we use our space flight capabilities to guard against such events in the future (see my essay "The Watch in Space" in *Change!*).

NOTE: *This essay appeared in March 1981. Subsequent to its appearance, the suggestion that an asteroid strike had destroyed the dinosaurs lost favor. It seemed more likely that a comet could have done all the damage without quite the same cratering effect. The following essay appeared in July 1984, and in it I discuss the matter again.*

48 NEMESIS

In the previous essay, I discussed the possibility that a sizable asteroid (or, more likely, a comet) struck the Earth about 65 million years ago and sent up enough debris to block out sunlight for a lengthy period of time. This cooled and darkened our planet and killed off almost all earthly life, driving three-fourths of its then existing species to extinction, including all the dinosaurs. At the time the essay was written, the strike was thought to be one of those damaging events that might happen once in a long while but at unpredictable intervals. Since then, however, scientists have studied these episodes of extinction, these "Great Dyings," with particular care, and it begins to appear that they are not unpredictable accidents at all. They seem to happen regularly, every 26 million years or so.

Why should there be such regular Great Dyings? We don't know of anything here on Earth or out in space that takes place every 26 million years and is deadly to us. But something seems to be happening at those intervals, so scientists are trying to reason it out. One possible source of bombardment is the hundred thousand or more asteroids that circle the Sun in the asteroid belt lying between the orbits of Mars and Jupiter. What if some sizable object passed through the asteroid belt every 26 million years? Its gravitational pull would disrupt many asteroidal orbits and would send a considerable number of these small bodies plunging through the inner Solar System. A few of them would be bound to encounter Earth. And yet a sizable body passing through the asteroid belt every few million years would surely affect the orbits of the planets, including that of Earth itself, and there is no sign that any such thing has happened. Cross out the asteroid belt.

There is another belt of small bodies surrounding the Sun, however, one that astronomers have never actually seen. Back in 1950, a Dutch astronomer, Jan Hendrik Oort, suggested that about a light-year out in space there could be a hundred billion or so small, icy bodies slowly circling the Sun—remnants of the original dust cloud out of which the Solar System formed. Every once in a while, the distant gravitational fields of the nearer stars disturb some of these bodies and send them down among the planets, where they become visible as comets. This suggestion is now accepted by astronomers generally.

But maybe it isn't the stars alone that disturb this "Oort cloud" of distant frozen comets. Maybe it is some body that circles the Sun and that every 26 million years or so passes through or near the Oort cloud and, by gravitational disturbance, sends a large number of the comets careening through the planetary system, so that a few are bound to hit the Earth. This suggestion has been made recently by two American scientists, Daniel P. Whitmore and Albert A. Jackson. They suggest that the Sun is really part of a double star system. (Most stars are.)

The Sun's companion could be very small, so small that it would be a "substar" and would barely shine, or perhaps would not shine at all. Even if it did shine, it would be so distant that it would not be seen except through an excellent telescope, and even then it would go unnoticed as a companion. (Who would study so dim a star to see whether it was slowly moving through space?) Whitmore and Jackson suggest that the star might be no more than one-fourteenth the mass of the Sun, or about seventy times the mass of Jupiter. That would be enough to make it radiate infrared light but would probably not be enough to make it give off any detectable quantity of visible light. If its average distance from the Sun were 1.4

light-years (8 trillion miles, or a third of the way to the nearest star), then it would be circling the Sun once every 26 million years.

The star might not be traveling in a circle, however, but in an ellipse. At one end of the ellipse it might be two light-years from the sun and at the other end only 0.8 light-years from the Sun. Even at its closest approach to the Sun, it would be too far away to be seen or to affect us noticeably by its gravitational pull, and we would remain ignorant of its existence. Still, at its close approach it would pass through the thickest part of the Oort cloud and for a million years, large numbers of comets would streak among the planets, and a few might hit us, with disastrous results. There has been a suggestion, therefore, that the Sun's companion body be called Nemesis, after the Greek goddess of vengeful destruction.

It might be worthwhile, once we have sizable telescopes out in space, to search the sky for small sources of infrared radiation and see whether any of them moves slowly against the background of other stars. If one does, it might be Nemesis, the Sun's companion body. But there's no need for immediate worry, even if Nemesis does exist. The last period of considerable extinction was 11 million years ago, so we needn't expect another for about 15 million years.

NOTE: *In the year since the above essay was written, the enthusiasm for the Nemesis hypothesis has noticeably diminished, and the arguments over "mass extinctions" continue fiercely. Nevertheless, the discussion of the effects of a large strike upon the Earth have helped give rise to another controversial matter, which I take up in the next essay.*

49 THE DEADLY DUST

IN 1971, THE ROCKET PROBE Mariner 9 was placed in orbit about Mars so that it could map the Martian surface in detail. Unfortunately, Mars was experiencing a planetwide dust storm at that time, and dust in its upper atmosphere completely obscured its surface for a while. Slowly, after some months, the dust settled, and Mariner 9 completed its mission.

Carl Sagan and others, however, worked out the effect the dust must have had on Mars. Ordinarily, sunlight reaches a planetary surface and warms it. The lower atmosphere warms because of its contact with the surface, while the upper atmosphere remains cold. Warm air grows less dense and rises, so there is air circulation, which tends to even out temperature differences. A large quantity of dust in the upper atmosphere, though, absorbs heat and warms the air about it; the lower atmosphere remains cold. The warm air above doesn't sink, and the cold air below doesn't rise. There is little circulation, therefore, and the lower atmosphere remains cold until most or all of the dust settles.

In 1979, scientists began to speculate that an occasional collision of an asteroid or comet with Earth might kick up enough dust to cut off enough sunlight and lower the surface temperature long enough to wipe out a great part of the life on the planet (see the previous two essays). Is there anything that happens on Earth itself that could create such a catastrophe? Volcanoes, especially the explosive kind, can hurl vast quantities of dust into the upper atmosphere. So can huge forest fires. In both cases there is some cooling, and unusual weather can then be experienced. A recent volcanic eruption in southern Mexico seems to have caused this. Neither volcanoes nor forest fires, however, can do enough to produce true catastrophe—only inconvenience.

In 1983, however, a number of scientists (of whom Carl Sagan is the best known) began to calculate the results of what might happen if there was a thermonuclear war. Everyone knows that exploding hydrogen bombs would kill vast numbers of people by the force of the bombs' blast and the fires they would start and the radioactivity they would produce. Some people have calculated that such a thermonuclear war would kill a billion people and would wound (to almost sure death) a billion more.

Killing half the people on Earth is an unbearable thought, but it suggests that at least the other half would remain alive to rebuild. Apparently not! Every thermonuclear bomb would kick a huge supply of dust into the upper atmosphere. If the warring nations, in their desperation, exploded most or all of their tens of thousands of nuclear warheads, the amount of dust in the upper atmosphere would reach colossal proportions. Add to this the fact that the bombs would set up vast fires in cities and forests and that these would add further dust to the supply. Even a quite small exchange of bombs would be enough (with the forest fires they would start) to fill the upper atmosphere with a dangerous quantity of dust.

So far, the calculations made by scientists in the United States, the Soviet Union, and Europe tend to agree. Sunlight reaching Earth's surface would be cut down to 1 or 2 percent of normal, so there would be a prolonged night lasting several weeks. Temperatures on the surface would drop to as low as thirteen degrees below zero, Fahrenheit, and stay below freezing for months (even if the war took place in summer). This would occur mostly in the northern hemisphere, where the war would be enacted, but 90 percent of the Earth's population lives there. What's more, the dust would be driven across the equator, and the Southern Hemisphere would experience some, if not all, of the bad effects.

The prolonged darkness and frost all over the world (the so-called nuclear winter) would kill plant life in vast quantities, and most of the animals (including human beings) who might have survived the blast, fire, and radioactivity would freeze and starve. By the time the dust settled and warmth returned, the Sun would be shining on a mostly dead world. Even if a few starving human beings had managed to survive, civilization would certainly be destroyed.

Of course, the calculations may be incorrect. Some factors may not have been taken into account, either because they were overlooked or because they are unknown, but who dares fight a nuclear war on the chance that the calculations are wrong and that we will kill only 50 percent of Earth's population rather than 99 percent? More than ever, it seems certain that nuclear war would mean the end for the "loser," the "winner," and the neutrals alike. It would be the end for *everyone*.

But there is hope in all this. The prospect of a nuclear winter may well make all nations understand that there must be no nuclear war under any circumstances. It may mean that, driven by an ever hardening public opinion, nations will be forced, even against their will, to strive for understanding and cooperation among themselves.

50 TOURING EARTH

THERE'S A GOOD CHANCE that during the twenty-first century there may be millions of human beings living in numerous space settlements in orbit about the Earth. At first, the people living in these settlements will have come there from Earth, and every one of them will, at one time or another, feel homesick for Earth. Some will experience the homesickness more deeply than others and, over the years, there will probably be a continual drizzle of returnees.

The settlements will make themselves as Earth-like as possible, we can be reasonably sure. The Sun will be reflected by a large mirror through Venetian blind arrangements that can be opened and shut to create day and night. The settlements will be made to rotate to produce a centrifugal effect that will mimic Earth's gravity. There will be plants and animals from Earth, architecture to give the impression of a cultural and technological background suitable for the population, and so on. Nevertheless, whatever the settlers do, the space settlements will be *small* worlds, and some settlers will be unable to endure the claustrophobia and will long for the presence of strangers and the existence of sprawl.

Still, there will be those who will stick it out and become out-and-out Settlers, perfectly content to dwell away from Earth all their lives. More important, children will be born on the settlements who will have never known any other way of life. It will be completely home to them. They will be *natives;* they will not know what it is like to live on Earth; and they may give themselves a special name—Spacers, perhaps.

And will Spacers never visit Earth? After all, however much they may think of a particular settlement as home and of other settlements as other "cities" in space, there will always be the vast globe of Earth hanging in their sky. There will always be television communication to make the Spacers aware of this strange world, enormously large and culturally complex, from which their ancestors came.

There would surely be a flood of Spacers visiting "the old country" at all times, savoring its infinite spaces, its complex ecological balance, the impossible extremes of its scenery. Settlements may have small animals and useful plants, but they are not likely to have giraffes, elephants, and redwood trees. They may have ornamental lakes, but none of them will have an ocean. They may have pretty hills, but none of them will have a mountain range. None will have a sky such as Earth's sky, and no amount of mere description or glimpses on a television screen will serve as substitutes for those things.

Visits to Earth won't be easy. For one thing, they will undoubtedly be expensive. For another, they will be most inconvenient, for returning to the settlement will probably mean subjecting oneself to a painstaking quarantine until the settlement authorities have been convinced that no undesirable plants, animals, or parasites will be brought back from Earth — or infectious diseases, either. Nevertheless, I can well imagine that the visit will be worth all the money and inconvenience it will involve.

But do you know what I think will be Earth's *greatest* attraction? Weather! The small settlements in space are bound to be designed in such a way as to have a pleasant and equable climate. There would be no temperature extremes and no meteorological violence of any kind. There would be no true rain, let alone hurricanes or earthquakes or tornadoes or anything of that sort. To be sure, settlements might be exposed to their own dangers, such as meteor strikes, but these would probably occur only rarely.

Anyone, therefore, who had lived on a settlement all his or her life might find nothing on Earth so fascinating as a heavy, drenching rain — all that water just falling down from above! Even better would be a snowstorm. Or the feel of wind in one's face — real gusts that would make it difficult to walk upwind. Even a heat wave or a cold wave might have a refreshing novelty to it, at least at first. Imagine what a Spacer would make of snowball fights or bellywhopping on a sled or skiing; or sitting about on a beach with little or nothing on; or traveling by ship on the ocean; or being someplace where there was no sign of land anywhere, especially with a fresh wind whipping the surface of the sea into froth—until he got seasick, of course.

Somehow I imagine that Spacer adults would enjoy all this the most. I'm not sure about the youngsters. It might be *too* novel for them. If you were a child who had never seen snow, might it not frighten you to feel bits of cold white stuff falling on your face? I suspect that Spacer youngsters would have to be introduced gently to the strange sights they would see on Earth—and maybe even some of the adults might profit by indoctrination courses; but I am sure it would all be worth it.

51 GETTING THE LEAD OUT

BACK IN 1859, a French physicist, Gaston Plante, devised a new kind of electric battery—a new way of setting up a chemical reaction that would produce an electric charge that could, in turn, be drawn off as a steady electric current. He placed one rod containing lead and another containing lead dioxide into a vat of sulfuric acid. Both the lead and the lead dioxide reacted with the sulfuric acid to produce lead sulfate, and in the process an electric charge was built up and an electric current could be drawn off.

What made this battery different was that it could be reversed! If too much electricity was drawn off—if too much lead and lead dioxide were converted into lead sulfate—an electric current could be forced through the battery in the opposite direction. The lead sulfate would change back into lead on one side and lead dioxide on the other side, and the battery could then be used all over again. Not only could the battery produce electric currents; it could also *store* the energy of an electric current, and it was therefore the first "storage battery."

That was a century and a quarter ago, and the storage battery used in your automobile to get it started and to run all the electrical accessories is, with only minor improvements, *still* the same storage battery that Plante invented. A few other storage batteries have been devised using nickel or silver in place of lead, along with iron, cadmium, or zinc. There are advantages and disadvantages to each, but the bottom line is that we still use the "lead-acid storage battery" more than any other.

Lead, however, is a dense metal, and if you've tried lifting an automobile storage battery, you know it is heavy. What's more, sulfuric acid is a very corrosive chemical, and it would be no fun whatsoever if you were to splash some on yourself. Fortunately, there is only one storage battery to an automobile; but suppose you wanted an electric automobile that would be silent, would not produce fumes, and would save fuel. You would need bigger and more numerous batteries—more weight and more acid. What's more, although the batteries could be recharged, it would not be a very rapid process.

It is time, after a century and a quarter, for some breakthrough, and there may be one on the way that will help us get the lead out of the automobile. In fact, we can get all the metal out of the battery. In many ways, the twentieth century has seen plastics replace older materials, and this may well prove true for batteries as well. Chemists at the University of Pennsylvania have been working with polyacetylene, which consists of a long chain of carbon atoms, to each of which a hydrogen atom is attached. It is the simplest carbon-chain plastic and was the first one found to conduct an electric current, at least to some extent.

The chemists found that if two sheets of polyacetylene are placed into a solution of a chemical called lithium perchlorate, the sending of an electric current through the sheets and solution will produce an interesting result. The lithium portion of the chemical goes into one sheet, and the perchlorate portion goes into the other. In this way, the two sheets store electric energy. When the sheets are then hooked up to an electrical system, the lithium and the perchlorate flow out of the sheets and an electric current is set up. What we have is a "plastic storage battery" that can discharge and then be recharged, over and over again.

As it happens, a plastic battery is much lighter than the customary lead-acid battery. If you were to build a plastic battery that was the same size and shape of the usual lead-acid battery, it couldn't store as much electricity. If you made a large plastic battery, however, that *weighed* as much as a lead-acid battery, it would store three times as much electricity. What's more, the plastic battery would contain no corrosive chemicals.

The fact that plastic batteries would take up more room than lead-acid batteries would not be as bad as it sounds. The plastic sheets could be molded into any shape, and they could be fitted into any part of the automobile. The car's roof or doors or hood could be hollowed out and then filled with an appropriately shaped plastic battery made out of very cheap material. Damage to the frame of the car would release no poisonous or dangerous chemicals, and the cost of replacing the batteries (if necessary) would be far lower than that of beating out the dents.

There are still bugs, to be sure. The plastic sheets deteriorate in air, and that, of course, is a fatal problem. However, it is possible to get around that obstacle, and better, more stable varieties of plastics may yet be found. In any case, the breakthrough may be on its way, and it may not be more than a decade or so before we can finally get the lead out. It's about time, too.

NOTE: *After the above essay appeared in November 1982, I received a letter from Werner T. Meyer of the Lead Industries Association which served to educate me a bit. He pointed out that the lead-acid battery has undergone considerable improvement even over the past two decades. The use of plastics in the battery has decreased its weight by 30 percent, and its length of life has increased to the point where a four-year use is not unusual. (When I got my first car, 1½ years was all I could count on for the battery.) Furthermore, lead can be recovered from old batteries with great efficiency. Finally, Mr. Meyer is very skeptical about the possibility that a plastic battery will actually come to pass in the reasonably near future. It seems to me only fair to let the reader know all this.*

---

52 SPLITTING WATER

HYDROGEN (H), WHEN BURNED in air, combines with oxygen (O) to form water (H_2O). In doing so, it yields a hot flame —which means a great deal of energy. If we could have all the hydrogen we wanted, that would solve our energy problems forever. Why not break up the water molecules back into hydrogen and oxygen, then reuse the hydrogen, and continue to do this over and over again? Unfortunately, just as combining hydrogen and oxygen with water *yields* energy, breaking up water into hydrogen and oxygen *consumes* energy. The water molecules must be heated to very high temperatures or else they must be subjected to an electric current under certain conditions. The best way of getting the heat, or the electric current, is to burn a lot of fuel and thus spend a lot of energy in forming the hydrogen.

As it turns out, the energy we spend to get the hydrogen is invariably greater than the energy we get back by burning the hydrogen we form. We suffer a net loss of energy every time.

(That's one aspect of the famous second law of thermodynamics.) However, green plants split the water molecule by using the energy of sunlight, not the energy of burning fuel or of the electric current. The water molecule splits into oxygen and hydrogen. The oxygen is released into the air (thus replacing the quantities used up when animals, including human beings, breathe). The hydrogen is combined with carbon dioxide from the air (which animals, including human beings, produce) to form starch, sugars, and, by adding minerals from the soil, proteins and all other components of plant tissue. In short, the hydrogen and carbon dioxide are used to form what we consider food. It is that food which supplies the chemical energy that keeps us (and all other animals) going.

The second law of thermodynamics still holds true, however. The energy of the sunlight that is used by the plants is greater than the energy they obtain from the hydrogen they produce; and that energy, again, is greater than the energy of the food they eventually form. Plants operate at a net energy loss, *but* it doesn't matter. The sunlight, as far as plants are concerned, is free and is always there. If the plants didn't use it, it would just go to waste. And sunlight will continue to be there, used or not used, for billions of years. So even while operating at a net loss, plants can continue to produce food for all animal life indefinitely.

What's more, the energy stored by the plant world in its tissues can be used as energy in other ways. Wood can be burned; coal is just fossilized wood. Animals, feeding on plants, store chemical energy as fats and oils; petroleum is just fossilized fat. The plant world supplies us not only with food but with wood, coal, and oil as well—all of its energy sources.

Can human beings duplicate the trick of the green plant? We haven't been able to so far. Plants are made up of cells that contain chlorophyll (a green substance that gives plants their characteristic color). Molecules of chlorophyll are teamed up with many other molecules in an extraordinarily complex system that can snatch the very dilute and low-grade energy of sunlight, concentrate it, and use it to break up the water molecule. Chlorophyll won't work by itself, and we don't know how to put it together with other substances into a working system like the one that exists within the plant cell.

But is it possible that substances other than chlorophyll might be able to perform the same trick of seizing the dilute, low-grade energy of sunlight and of using it to split the molecule of water? That would give us hydrogen, which we might either use as such or combine with carbon dioxide or other substances to form natural gas, gasoline, and so on. What's more, substances formed in this way would be without measurable amounts of impurities, so there would be no chemical pollution formed on burning. Even the carbon dioxide and heat that are formed on burning would merely be replacing the carbon dioxide and heat used to form the various fuels in the first place. Not only that, but the hydrogen and carbon dioxide (plus other easily available substances) could be used to form plastics, fibers, medicinals, and any number of other useful substances. All we need is a chlorophyll substitute.

And chemists are on the track! Melvin Calvin of the University of California is one of the renowned experts in the field of chlorophyll reactions (he's won a Nobel Prize for his work), and he is using synthetic, metal-containing compounds designed to mimic the activity of chlorophyll. Others are also working in this field. So far, no one has quite created the equivalent of an artificial plant cell, but there is no reason why success should not come about eventually, and perhaps even in the fairly short run. Then we might produce hydrogen out of the cheapest and most common things on Earth—water and sunlight—and this might prove to be the best way of making use of solar energy, possibly even in space.

53 GOSSAMER WINGS

ONE OF THE DREAMS of would-be space engineers these days is to design and build huge solar-powered satellites that would orbit around the Earth. Such satellites would include a broad array of "solar cells" that could convert sunlight into electricity. The electricity would be converted into radio waves and would then be beamed to Earth, where they would be received and converted back into electricity again.

It wouldn't be easy. In order for the satellites to absorb enough sunlight to produce an amount of electricity that would be significant in terms of how much the planet uses, perhaps a hundred of them would have to be spaced along an equatorial orbit about twenty-two thousand miles above Earth's surface. (That would place them about one thousand six hundred miles apart.) Each satellite would have a mass of about fifty thousand tons and would have to expose an area of solar cells equal to that of the island of Manhattan. And each one would cost at least $20 billion.

Of course, once such a system were completed and working, the world would be basking in ample Sun-borne energy, and there would be no need to fear a shortage for billions of years. We might all then agree the effort had been worthwhile. Getting from here to there, from nothing to a hundred orbiting satellites, is a problem, however, and a big one.

Anything we could do to cut down the mass and expense of such satellites would naturally put us that much further ahead in the project, and one expense involves the solar cells. Right now, the best solar cells we can make are made up of silicon crystals. This doesn't sound bad, to be sure, since silicon is the second most plentiful element on Earth. Nearly one-quarter of the atoms in the Earth's crust are silicon atoms. These silicon atoms, however, must be separated from the rocks in which they are found, and must be highly purified. They must then be so treated as to form single, large crystals. All this is tedious and difficult work, and by the time the work has been completed, the cost of obtaining electricity from such carefully made material is anywhere from three to ten times higher than the cost of obtaining it in the conventional manner.

Scientists are working out ways, though, of making use of "amorphous" silicon. In crystals, all the silicon atoms are present in neat, endlessly repeated arrangements. In amorphous silicon, on the other hand, the silicon atoms are arranged any which way. Clearly, amorphous silicon is far cheaper to prepare than the crystals are, so there would be an advantage to using the amorphous stuff even if it were somewhat less efficient in converting light than the crystals were.

The use of amorphous silicon can cut down mass as well. When silicon crystals are used, a large, thick crystal is formed and is then sliced very thin. The solar cells formed out of the thin slices are very light individually, but the mass would mount up enormously, considering that every solar power satellite would have to present no fewer than a billion cells to the sun in order to catch enough light to be useful. If, however, scientists do manage to work out the techniques for using amorphous silicon, they may well be able to layer it onto some suitable surface in such a way as to form a very thin film, one that is much thinner than even the thinnest practical slice of silicon crystal. The mass of amorphous silicon film would be perhaps a tenth of one percent of the mass of the slices of silicon crystal that would serve to cover the same area. Economy of manufacture would be further enhanced by layering the amorphous silicon onto a continuous roll of material, with the silicon then etched into small units.

What would be the material onto which the silicon would be layered? If the material were itself massive, of what use would it be for the silicon to be light? Suppose, though, that the amorphous silicon were layered onto very thin plastic. The whole thing would be very cobwebby, and we could end up with the "gossamer wings" Cole Porter talked of in his ballad "Just One of Those Things." Porter's gossamer wings took lovers to the Moon, but these would bring the Sun to us—an even more glamorous effect, in my opinion.

There are other materials besides silicon that might serve, too, and other techniques. All in all, it seems reasonable to suppose that in the next few years both the price and mass of solar cells will be substantially reduced. This means that although solar-powered satellites will continue to be formidable undertakings, they will begin to seem just a bit less formidable. Surely, there will soon come a time when the benefits to be gained will warrant the investment and risk of construction.

You might wonder, by the way, whether the gossamer wings I describe can stand up to the work expected of them. Stretching flimsy material over a unit the size of Manhattan wouldn't seem to offer a very durable system. We must remember, however, that there is no weather in space; no winds, no rain, no ice, and, moreover, no vandals. Even gossamer would hold up—except for one thing. There are dust particles in space, even pebbles, very occasionally. The wings will be etched, even punctured from time to time, and repairs will now and then be required.

54 ALL THE MASS

ENERGY AND MASS are interrelated. If something gives off energy, it loses mass; if it absorbs energy, it gains mass. However, it takes so much energy to produce so little mass that throughout most of human history, no one noticed the interrelationship. In fact, throughout the 1800s, scientists were convinced that mass was never either lost or gained. Consider a quantity of gasoline that is burning, for instance. In human terms, that produces a lot of energy, enough energy to send an airplane speeding through the air at a thousand or more miles an hour. And yet when gasoline is burned, only about a two-billionth of its mass is lost. That is enough to produce the energy that is released during the burning, and it is no wonder that so small a loss went unnoticed.

Albert Einstein, in 1905, worked out the mass-energy interrelationship from theoretical considerations. That immediately explained where the energy of radioactivity came from. A substance such as radium (discovered seven years earlier) produced so much energy for every atom breaking down that no previously known source of energy would have sufficed to explain it.

Ordinary burning, such as that of gasoline, involves only the tiny electrons on the outskirts of the atom. Radioactivity involves the much more massive particles in the nucleus that is at the very center of the atom. Radioactivity is an example of *nuclear* energy. Humanity first encountered nuclear energy in large doses with the invention of the nuclear bomb, one that involved uranium fission (that is, the breakdown of the large uranium atom into smaller atoms).

When a quantity of uranium undergoes fission, about 1/700 of its mass is lost. This is still not much, but it is about two and one-half million times the amount of mass lost when gasoline burns. This means that if a certain mass of uranium undergoes fission, it produces 2.5 million times as much energy as that same mass of gasoline burning. That is why scientists have been anxious to bend uranium fission to peaceful uses, and why they have labored to build safe nuclear power plants (not an easy job).

And yet fission is not the ultimate. When a quantity of hydrogen (made up of the smallest known atoms) undergoes fusion to larger atoms, it loses about 1/110 of its mass. This means that a pound of hydrogen undergoing *fusion* will produce about 6½ times as much energy as will a pound of uranium undergoing *fission*. Nuclear bombs that involve fusion are called hydrogen bombs, and it is no wonder that they are far more destructive than were the original bombs, which involved only fission.

So far, we have not been able to produce controlled fusion that can be turned to peaceful uses. There is every sign that we will eventually do so, however, and once we have fusion power stations that (it may well prove) are less dangerous than fission power stations, humanity's energy problems may be solved for the foreseeable future. But fusion isn't the ultimate, either. Even after fusion, more than 99 percent of the mass is still there. Is there any way of using *all* the mass?

Yes, there is. Every subatomic particle has an "anti-particle" that is its equal, but opposite. About 99.9 percent of the mass of a hydrogen atom is a proton located at the very center of the atom. Matching it, there is such a thing as an "anti-proton," out of which an "anti-hydrogen atom" can be formed.

If a proton and an anti-proton should happen to meet, they undergo "mutual annihilation." The *total* mass of the two particles vanishes and is replaced by the equivalent amount of energy. If half a pound of protons and half a pound of anti-protons were to meet and undergo mutual annihilation, they would produce 110 times as much energy as a pound of hydrogen undergoing fusion. This would be the ultimate, but the trouble is that there are virtually no anti-protons in nature. The universe around us is full of protons but not of their opposite.

Scientists have learned, fortunately, how to produce anti-protons. Once this has been done, however, there is the even greater problem of storing them, for the moment anti-protons touch ordinary matter, they encounter protons, which are to be found in all matter, and they are then promptly annihilated. Scientists are therefore learning how to suspend anti-protons in a vacuum through the use of magnetic fields and by other methods that would keep them from ever touching matter.

This could probably be done more easily in outer space than on Earth, so the time may come when there will be special space stations devoted to the production and storage of anti-protons. These will be supplied, as fuel, to spaceships also built in space. Powered by proton/anti-proton mutual annihilation, spaceships would be able to move faster and have a greater range than anything we can possibly build now. With them, we may even aim for the stars.

55 HOTTER THAN HOT

FOR YEARS NOW, physicists have been trying to work out the details of the "big bang" with which the universe started. The big bang itself is time-zero, but what was the universe like one second after it—one-hundredth of a second after—one-millionth of a second after—one-trillionth of a trillionth of a second after?

Physicists have worked out incredibly ingenious scenarios based on what they understand of the laws governing the behavior of subatomic particles, but there's something missing. The farther back they try to penetrate into the beginning—the closer they try to get to the big bang—the tinier the universe becomes and the higher the temperatures that they have to deal with. The very early universe was incredibly more hot than hot—so hot that physicists don't really have any data that begin to apply. It's a matter of trillions of trillions of trillions of degrees, and physicists aren't even sure that the laws of nature, as we know them, don't break down altogether at such temperatures. If only they could study the behavior of matter at higher temperatures than they have so far been able to reach, they might feel a little more confident in their reasoning.

One way of obtaining very high energies (and, therefore, very high temperatures) is to begin with some tiny object that carries an electric charge. Such a charged object can be accelerated in a magnetic field and made to go faster and faster. Eventually, it is smashed into a metal target. From the results of the collision, one can work out some details of the basic structure of matter, the laws governing the most fundamental aspects of the universe, and so on. The more massive the particle in question, the more energy is developed, and the most massive particle easily available in quantity is the proton. Devices for accelerating protons have been made more and more powerful over the years, and in the 1970s, twin devices came to be used that accelerated protons in opposite directions. Protons were then smashed, not into stationary targets, but into each other in head-on collisions. Much more energy was released.

The next step would be to use something more energetic than a proton. Every atom has a central nucleus with an electric charge. The hydrogen atom has the simplest nucleus, just a proton. Other more complicated atoms have nuclei made up of a number of protons together with a number of neutrons. (Neutrons are as massive as protons but have no electric charge.) The most complicated atom that is found in nature in quantity is uranium. Its atomic nucleus has 92 protons and 146 neutrons. Altogether, a uranium atom is 238 times as massive as a single proton.

However, every atomic nucleus is surrounded by light electrons that tend to shield it and to neutralize its electric charge. To make it possible to accelerate these nuclei, the electrons have to be removed. It is easy to remove the one electron that a hydrogen atom has and to bare the single proton at its center. More complicated atoms have more electrons, and it is always easy to remove one or two. The more electrons that are removed, however, the more difficult is the removal of the remaining ones.

In 1984, at the Lawrence Berkeley Laboratory of the University of California, some startling progress was made. A few electrons were removed from uranium atoms. This gave each uranium atom an electric charge, so that magnetic fields could accelerate it, making it move faster and faster. At some suitable speed, the uranium atoms were sent into a copper foil about as thick as a sheet of paper. In passing through and shouldering their way between the copper atoms in the foil, the uranium atoms had some of their electrons scraped off, so to speak. The faster the uranium atoms move, the more electrons are scraped off. When the uranium atoms are made to move at seven-eighths the speed of light (160,000 miles per second), then 85 percent of the atoms shoving through the copper foil emerge without a single one of the 92 electrons that originally surrounded the nucleus. The other 15 percent have but a single electron remaining.

If such very massive nuclei with few electrons are carefully studied, a new and better view of the behavior of subatomic particles may be gathered. What is even more interesting, however, is that these bare uranium nuclei can be treated as though they were very, very massive protons. They can be accelerated in opposite directions and made to smash into each other. At any given speed, they could develop up to 238 times the energies that simple protons would. They could achieve, momentarily at least, temperatures and pressures far higher than any that physicists have yet observed in the laboratory. This would give physicists useful data, actually observed and not just calculated, in areas of unprecedentedly high temperatures, and they might be able to refine their knowledge of the very early universe in light of that—and better understand, in consequence, the nature of the universe today.

56 THE WEAKEST WAVES

THE EARTH'S GRAVITATIONAL PULL is something human beings have been aware of from earliest times. Every time we drop something or lift something, every time we fall, we are aware of it. Yet gravitational pull is the weakest force in nature by far. Electromagnetic attraction (the force holding a proton and an electron together inside an atom) is several thousand trillion trillion trillion times as stong as the gravitational attraction between those same particles. Why, then, are we so aware of gravitational attraction and so unaware of all the electromagnetic attractions in the atoms all about us and within us?

The answer is that electromagnetism involves both attractions and repulsions, and they occur in roughly equal amounts in ordinary matter. The two cancel each other out and leave us unaware of either electromagnetic attraction or repulsion. Gravitation, on the other hand, involves only attraction, so the more matter there is, the more attraction. Even so, the gravitational attraction is so weak that it takes a really large amount of matter to make it noticeable. A whole mountain range barely exerts a measurable gravitational pull. The only reason we're so aware of gravity here on Earth's surface is that the Earth is far, far larger and more massive than a mere mountain range.

According to Albert Einstein's general theory of relativity, the effect of a gravitational pull is produced by the way that mass distorts space in its vicinity. Space itself curves about the mass, and the motion of any object follows the curve, so the Earth as it moves curves about the Sun. Also according to Einstein, if any mass were to accelerate its motion—speed up, slow down, or change its direction—it would lose energy by giving off gravitational waves. The Earth accelerates because it constantly changes direction as it circles the Sun, so it gives off gravitational waves, loses energy, and approaches the Sun. However, the gravitational force is so weak that gravitational waves are trillions of trillions of trillions of times weaker than electromagnetic waves, such as those of light. In Earth's whole history, not enough energy has been lost through its radiation of gravitational waves to matter.

Naturally, the larger the gravitational acceleration, the more energetic the gravitational waves, but even the most energetic gravitational waves are incredibly weak. Einstein thought the gravitational waves were there but were too weak ever to be detected. Nevertheless, for some twenty years now scientists have been trying to detect gravitational waves. Since the gravitational effect depends upon the distortion of space and time, a gravitational wave produces a quiver in that distortion. Any object, even the Earth itself, quivers as a gravitational wave passes over it.

It's not much of a quiver. A large aluminum cylinder might quiver in and out by a thousandth of the width of a single proton —and yet that quiver could be detected by the proper electronic devices. The difficulty would be to isolate the tiny quiver from other effects that might confuse things. You might have to bury the cylinder in a deep mine to keep off the effects of radiation or keep it in a vacuum to avoid air currents or keep it close to absolute zero to negate the effects of heat. Moreover, you would have to have several such cylinders at widely different spots. A gravitational wave passes over the entire Earth as though our planet were only a dot in space. That means that all the various cylinders would have to show a quiver *at the same time* as the gravitational wave passed.

So far, gravitational waves have not been clearly detected, but there are plans for still more sensitive devices, and there are even dreams of setting up something in space that might detect such things more sensitively than we can here on Earth. Why bother? Well, it is only in the last quarter century that astronomers, through studies of cosmic radio waves and x rays, have come to realize what a violent place the universe is. If we could detect gravitational waves, it would be only the most energetic, and these would give us an idea of the most violent events of all.

Such gravitational waves as we could detect would be from very violent supernovas and from neutron stars revolving about each other at close range and, losing gravitational energy quickly, eventually smashing into each other. We might detect the gravitational waves given off when a giant black hole gulped down an entire star or, most of all, when two black holes struck each other and fused. All this might give us just the kind of information we need to understand more clearly how the universe started, how it developed, and how it may end. These are matters, after all, which scholars have pondered since the dawn of civilization, and anything that gets us even just a bit closer to an answer is supremely exciting.

---

57 A MIRROR IN PIECES

NEARLY SEVEN YEARS AGO, in my essay "Eye in a Vacuum" (see *Change!*, Houghton Mifflin Co., 1981), I discussed plans to place a large telescope in space. The telescope is not yet there, but the plans still exist and we can hope that it won't be too long now before they are realized.

Though such a telescope ought to yield marvelous results,

there are problems. It will have to be controlled at long distance and any information will have to be sent long distance. If anything goes wrong, it will take a long time to make repairs. Is it possible, then, while making use of the Space Telescope, to build one on Earth's surface as well, one that will be as good as or better than the one in orbit? It may not seem possible. Back in 1896, a refracting telescope was built with a 40-inch (1.02-meter) lens. That's about the biggest lens that can be practically ground into perfect shape and kept from distorting under its own weight as well as from absorbing too much of the light that passes through. After nearly a century, there is no larger telescopic lens.

Mirrors can be built larger than lenses for a variety of reasons. In 1948, the 200-inch (5.08-meter) mirror was placed in the telescope on Mount Palomar in California. It is still the best telescope of its type in the world. Larger ones have been built, notably one with a 236-inch (6-meter) mirror in the Soviet Union — but that one doesn't work very well. Bigger mirrors are too big and too heavy to stay rigid, too hard to grind perfectly, take too long to cool evenly, and so on. However, smaller mirrors will do the work if a number of them are used together. In 1976, a telescope was put into action in Arizona that possessed *six* 72-inch (1.83-meter) mirrors all working in unison. They are equivalent in light-gathering power to a single 176-inch (4.5-meter) mirror, but the six 72-inch mirrors are much easier to grind and to keep rigid than a single 176-inch mirror would be.

Well, then, why didn't astronomers think of that before? It wouldn't have done them any good if they had. The six mirrors have to work in unison with enormous precision. No human hand or eye could manage that precision, so such a multiple-mirror telescope had to await the development of adequate computers to do the job. Even so, the multiple mirror is not as good at resolution (seeing very fine detail) as an equivalent single large mirror would be. Astronomers are therefore preparing a new attack on the problem.

At the California Institute of Technology, there are plans to begin work on a new kind of telescope. It will have a single mirror—but in pieces. There will be thirty-six hexagonal (six-sided) pieces, each one 71 inches (1.8 meters) across. They will all be fitted together so perfectly that they will make up a single 394-inch (10-meter) mirror. Such a mirror would have four times the light-gathering power of the 200-inch mirror at Mount Palomar and would be much better in resolution, too. The advantage to such a mirror in pieces is that each separate piece, being relatively small, would be easier to cast in the first place and would take less time to cool down into an unflawed piece of glass. Grinding each piece into a perfect shape for precision focusing would not be easy, but it would be far easier to do it thirty-six times than to do a 394-inch mirror once. Finally, each piece, thanks to its comparative smallness, would not weigh much and would need to be less thick to be adequately rigid. All of the pieces put together, despite being twice as wide as the Palomar mirror, would weigh only one-third as much.

The catch is that the thirty-six pieces will have to fit together with a deviation of not much more than one ten-millionth of an inch. Since temperature change, wind, and gravity will tend to knock the pieces slightly out of alignment, astronomers will have to attach pistons to each segment and have them adjust their positions three hundred times per second under the guidance of a computer. The plan, moreover, is to put the telescope containing this segmented mirror on Mauna Kea in Hawaii, at a height of 4,200 meters (2.6 miles). At that height, three-fifths of the atmosphere will be beneath the telescope, and it will be that three-fifths that will contain most of the fog, moisture, temperature change, and quivering that plague astronomers.

It will be terribly cold up there, to be sure, and there will be so little oxygen that there will have to be sealed rooms containing higher air pressure where astronomers can catch their breath periodically. However, it will all be worth it to them.

And if the telescope works, there are hopes that a second one exactly like it might be built 100 meters (328 feet) away, so that the two together, used in tandem, would have ten times as fine a resolution as either one, working separately, would have. Who knows what we'll find out about the universe with tools like that?

58 SHARPENING THE FOCUS

EVER SINCE HUMAN BEINGS appeared on Earth, they have been able to look at the night sky and see light sources, from the full Moon down to very faint stars. More than three and a half centuries ago, telescopes came into use, and astronomers could gather more light than with the unaided eye alone, so they were able to make out fainter stars. In the 1800s, they learned how to use the spectroscope to analyze the light waves and gather information from the different wavelengths that were present or absent. They learned to use cameras to freeze the light waves on film. And they built larger and larger telescopes, until we had one with a mirror 200 inches across, in 1948. But always it was light, light, light—

In 1931, it was discovered that short radio waves (microwaves) also reached us from the sky. They couldn't be seen by the eye, but could be received by appropriate instruments. Unfortunately, the necessary instruments didn't exist, but during World War II, instruments for handling radar were developed under the intense pressure of overwhelming necessity. Radar made use of microwaves and, through the instruments developed in this way, "radio astronomy" came into existence after the war.

Large "radio telescopes" that looked like huge, round, shallow dishes caught the microwaves and reflected them to a focus, where the instruments for recording and analyzing them were placed. The microwaves were emitted by energetic objects whose light was otherwise drowned in trillions of other light sources. What's more, the microwaves penetrated dust clouds that obscured light. Microwave emissions could even be detected in the daytime. By studying microwaves, astronomers discovered such things as quasars, pulsars, black holes, exploding galaxies, and the active centers of ordinary galaxies, including our own, which could never have been detected by ordinary light.

There was a catch, though. Microwaves were a million times as long as light waves, so they could show things only fuzzily. The long microwaves "stepped over" tiny things. Observing the universe by microwaves was like looking at a light through frosted glass. Of course, we could make up for that by using a larger dish. The larger the dish, the more clearly we could see. However, since microwaves are a million times as long as light waves, dishes are required that are a million times as wide as ordinary light telescopes. For a radio dish to "see" as sharply as the 200-inch telescope at Palomar, it should be 200 million inches across. That means it would be 3,157 miles across and would have twice the area of the United States. This is obviously impossible, and it would seem that we would never see clearly by microwave. Not so! We already do, and I'll tell you how.

It is not necessary to use a single enormously large dish. You simply use one sizable dish here and another sizable one there. If both dishes are timed by super-accurate "atomic clocks" and are made to move in unison by clever computerization, they behave like one dish stretching from here to there. Such dishes are said to be "long base line" and even "very long base line" radio telescopes. Cooperating dishes in California and Australia have produced a baseline of 6,600 miles. There are limits, of course. If we placed dishes on opposite sides of the Earth, we would have a base line of 7,900 miles, and that is the maximum —on Earth. What's more, most dishes on Earth are separated chiefly in the east-west direction rather than the north-south, and that limits certain observations.

Now, however, astronomers are beginning to think of establishing large radio telescopes in orbit around the Earth. There would be difficulties, of course, because dishes on Earth and in space would always be moving with respect to each other, but this could be taken into account. The first such dishes might be only three thousand miles or so above the Earth's surface and wouldn't do as well as the best Earth-bound systems, but they would help work out the feasibility of such things. With proper orbiting, they could produce north-south separations, too. Eventually, the hope is that a large radio telescope can be built in the hundred-thousand-mile range, or even on the Moon.

A Moon-dish working together with an Earth-dish would give a base line of 240,000 miles, and the Moon's orbit is known very well, so changes in direction and distance could be carefully allowed for. The focus in such a case could be sharpened indeed, for distant objects in the heavens could be seen with perhaps thirty times the detail that they could be seen with from Earth. What new discoveries would we make as a result; how much more would we learn about the universe generally—its origins, development, and possible end? We cannot predict what we would find out.

That is the exciting thing, actually. If we could predict the nature of new discoveries, why go to all the trouble of making them?

59 SAILING THE VOID

IN THE EARLY DAYS of civilization, human beings discovered the use of sails. Large areas of tough textile caught the wind, and the momentum of that moving air was transferred to the floating ship, which then moved without a water current, and even against one. With a wind in the sails, ships moved upstream without any need for the unending application of human muscle. Of course, the wind might not always blow, or, if it did, it might not blow in the right direction. Nevertheless, sail technology advanced steadily, so that feebler and feebler gusts could be made use of, even gusts that were in the wrong direction.

By the 1850s, the Yankee clippers were the speediest and most beautiful ships the world had ever seen. The steamships that replaced them were larger and, eventually, faster, but they were also uglier, dirtier, and noisier. Sailing ships (except for a relatively few pleasure vessels) have now disappeared from the world's oceans, but humanity faces another ocean today, an infinitely vaster and emptier one. Our water ocean stretches for tens of thousands of miles, but the ocean of outer space stretches for billions of trillions of miles.

We've begun the navigation of outer space with the equivalent of steamships—the use of raw power, incredibly noisy power. Nothing else will do, perhaps, to break through Earth's atmosphere and the lower reaches of its gravitational field. Once a ship is in space, however, and is moving through a vacuum in orbit about the Earth, is there anything quieter, gentler—better? A wind would be, certainly; but outer space is a vacuum. What is there in space that can form a wind?

Two things, actually—at least in the neighborhood of a star like our Sun. The Sun emits high-speed, electrically charged particles in a continuous stream outward from itself in all directions. This is referred to as the "solar wind." The solar wind possesses momentum, and this momentum can be transferred to anything that blocks it.

A comet, for instance, as it approaches the Sun, is partially vaporized. The rocky dust frozen in its outer layer then surrounds the still frozen nucleus in a haze. This dust is swept outward from the Sun by the solar wind to form the long "tail" that is the most spectacular part of the cometary vision. Of course, the solar wind is very rarefied, compared with Earth's atmospheric winds, and it can move only tiny particles, such as those of the cometary haze. It has no useful effect on ordinary spaceships. What's more, the force of the solar wind grows rapidly weaker as one moves away from the Sun and as the particles of the wind spread out more and more widely, with fewer and fewer of them striking the spaceship.

The second "wind" in space is an even less substantial one, for it is the Sun's light, which also possesses momentum, and which exerts a minuscule pressure. Light exerts less pressure than the solar wind does, and, like the solar wind, it grows weaker with increasing distance from the Sun. If the solar wind is not useful as a way of moving spaceships, why should light be? The answer is that we can manipulate light more easily than we can manipulate the charged particles of the solar wind. There are ways of converting the ordinary light of the Sun into a laser beam, which is a wave of "coherent" light. This is light in which all the waves are the same length and move in the same direction. Whereas ordinary sunlight spreads out rapidly, so that its pressure weakens into utter uselessness, a laser beam spreads out only very slightly, so its pressures, and its ability to move a ship, can remain constant over very long distances.

Even a laser beam exerts very little pressure, however, so a ship would have to have a way of intercepting a lot of it. The ship would have to be attached to a sail, in other words, and a very huge sail at that, one which would intercept a great deal of light. We would have to imagine a ship in the center of a sail made of very light, opaque material that would be something like two or three miles across. A laser beam from Earth could drive such a ship through space by solar energy, burning no fuel and never running out of energy. Such a ship could be driven to the nearest star, Alpha Centauri, in forty years.

It might seem that such a laser beam could drive the ship only outward, never to return; but with the use of still larger sails and ingenious ways of extending electrically charged systems, such ships could be brought to a halt in the neighborhood of Alpha Centauri and might even be made to take up the return journey. The requisite technology is a bit beyond us today, but a hundred years from now it may not be, and sailing vessels more magnificent than any we have ever seen may navigate distances vaster and emptier than anything our ancestors could have dreamed of.

60 MOON SPLASHES

IN MY ESSAY "The White Background" (see *Change!*, Houghton Mifflin Co., 1981), I described how much easier it is to locate meteorites on the featureless ice of Antarctica than on the broken, rocky ground that marks the other continents. I said that a serious search of Antarctica ought to be made to locate such meteorites, which are solid bits of matter that circled the Sun for billions of years before finally happening to reach a spot in space that Earth had also just happened to reach.

Scientists didn't really need my urging. They were already engaged in the search, and since that essay was published, over 6,000 meteoric bits and pieces have been located. The bits of matter that circle the sun are called meteoroids. Once they strike Earth's atmosphere, they dash through at a dozen miles a second or so, and in the process are heated to a white-hot glow. Some of the outer portion of the meteoroid melts and flakes off, and the part that survives to reach the Earth's surface is the meteorite.

Where did the meteoroids come from? Astronomers feel

confident that meteoroids originated in the asteroid zone, where the original matter out of which the planets were formed could not manage to coalesce properly because of the gravitational disturbance of the nearby giant planet, Jupiter. The effect of occasional collisions between asteroids is to send small fragments into orbits that take them closer to the Sun and into the neighborhood of the inner planets (including Earth). This means that the meteorites we pick up represent matter that solidified four and a half billion years ago and has remained almost unchanged ever since. Meteorites give us information concerning the early days of the Solar System, information we can't get from Earth's rocks. The Earth underwent vast early changes because of volcanic action, water and air erosion, and so on, so we can't find Earth rocks older than about 3.5 billion years. The first billion years are a blank.

There are other possible origins of meteoroids, too. In the early ages of the Solar System, the planets were still forming, and large chunks were striking still larger chunks and coalescing. The final strikes left craters—splash marks—that remain to this day on most medium-sized bodies in the Solar System. On Earth, these marks were erased by the action of water, air, and life, but on the Moon, which has none of these, visible craters remain after billions of years.

When a large meteorite strikes Earth and produces a splash, little if any of that splash can escape the Earth, for splashed bits must travel at seven miles a second to get away. If the Moon is struck and splashes, bits can escape at only 1.5 miles a second, since the Moon is a smaller body than the Earth and has a weaker gravitational pull. Some bits undoubtedly do escape and become meteoroids. Then, eventually, a few of these bits of Moon splashes strike the Earth. It seems likely that at least two of the Antarctica meteorites had their origin on the Moon. For one thing, thin slices of them have a makeup that can be explained by supposing they originated in a vast crater-forming collision. The makeup is not like that found in other meteorites, or in Earth rocks, but resembles very much the appearance of bits of Moon rock brought back to Earth from the neighborhood of craters. Then, too, the relative proportions of atoms of different metals present in the suspected moon splashes are quite different from those of other meteorites but are very similar to those in Moon rocks.

Finally, oxygen atoms exist in three different varieties, or isotopes. The proportions of these isotopes in the suspected Moon splashes are similar to those in some meteorites which are, however, quite different in microscopic appearance. They are also similar to Moon rocks, which are the *same* in microscopic appearance. If more such Moon splashes are found, it may be possible to deduce where on the Moon each has come from. If so, we would have Moon rock samples from places our astronauts have never visited. We might then have interesting and important information which, till now, we have had no other way of obtaining.

You may be wondering why we had to go to all the bother and expense of going to the Moon to collect rocks, when some of them were kind enough to come here. The explanation is that we would never have recognized a meteorite to be a moon splash if we hadn't first obtained Moon rocks to compare it to. Incidentally, one meteorite has been reported to contain certain atoms of gases in proportions that conform closely to the proportions of those atoms in the atmosphere of Mars. It may be possible that this meteorite is a "Mars splash." To be sure, Mars has a stronger gravitational pull than the Moon does and is much farther away, so very few Mars splashes are likely to reach us. But even if this meteorite is the only one from there, it's here, and one is better than none.

61 SKIMMING THE COMET

IN THE YEAR 1304, an Italian artist, Giotto di Bondone (best known simply by his first name, Giotto), completed a great painting, *The Adoration of the Magi*. In it the wise men are shown worshiping the newly born Jesus. The Star of Bethlehem is usually included above the manger in such paintings, often in a very stylized fashion. Giotto, however, presented it as a comet and painted it in a very realistic way. This was additionally unusual because comets, which were thought to presage disaster, sent people into a panic and were frequently pictured as swords or in even more dramatic and horrifying ways. Giotto, however, merely painted a comet.

As it happened, a very bright comet had appeared in the sky in the fall of 1301, and Giotto had probably observed it carefully with his painter's eye. He probably painted the comet in 1304 as he had seen the comet in 1301, which was the one we now know as Halley's comet, or (making use of a new system) Comet Halley. It is called Comet Halley because the first astronomer to work out its orbit was the Englishman Edmund Halley, in 1705. He demonstrated that it followed a long ellipse around the Sun and returned to the inner Solar System, where it could be visible to people on Earth every seventy-six years (give or take a couple). Its previous appearance had been in 1682, and counting backward five appearances from that one would take us to the comet Giotto saw.

We can also look forward to future visits of Comet Halley. Halley himself predicted its return in 1758, and it returned on schedule. Since then it appeared in 1835 (the year Mark Twain was born) and in 1910 (the year Mark Twain died). In 1986 it will be in the sky again. In fact, it has already been seen. It was spotted on October 20, 1982, but it was still farther away than Saturn at that time, and it could be seen only with the very best astronomical equipment. It will be a while before ordinary viewers can see it just by looking at the sky. Unfortunately, on this visit, it will pass Earth at a distance and will appear rather faint. It will hardly be seen at all in the Northern hemisphere.

Astronomers are excited, though, because rockets exist that can take a close look at Comet Halley for the first time. This is important because in 1950 the Dutch astronomer Jan Hendrik Oort suggested that comets existed in a huge spherical cloud one or two light-years from the Sun. In that same year, the American astronomer Fred Lawrence Whipple suggested that comets were made up of icy substances, with fine rocky dust or bits of gravel spread through them and with a rocky core, possibly, at the center. Far out in the comet cloud, the material is frozen solid. When one of the comets happens to be forced into an orbit that takes it into the inner Solar System, the heat of the sun vaporizes some of the ice, liberating the dust and gravel. This surrounds the icy "nucleus" with a cloudy "coma." The coma is swept away from the Sun, a little at a time, by the solar wind, forming a tail that can be very long if the comet is a large one and hasn't visited the Sun too many times.

It is possible that the comets in that far distant cloud, way beyond the planets, are made up of the contents of the huge cloud of gas and dust that condensed to form the Sun and its planets. This matter should be unchanged in all the billions of years since, and astronomers would like to know its composition. They can't go out far beyond Pluto and examine the comets in their original orbits, but they can study the occasional comet that wanders into our neighborhood. Such comets have been changed as a result of solar heating and evaporation, but they are better than nothing.

Over the course of the next year, therefore, Japan, the Soviet Union, and a combination of Western European nations (but *not* the United States) will each send out unmanned rocket probes to investigate Comet Halley as it passes through the inner Solar System. Of these probes, it is the European one that is scheduled to make the closest approach to Comet Halley. It will skim by it at a distance of about 310 miles (roughly the distance from New York to Montreal). It will be plunging through the comet's coma at a speed of about 45 miles per second and may be destroyed in the process. Before it is put out of action, however, it will analyze the chemical substances in the cloud and perhaps take photographs of the surface of Comet Halley's icy nucleus. My own guess is that it will completely confirm Fred Whipple's suggestion as to what comets are made of.

And what name do you suppose has been given to this adventurous probe? Why, Giotto, of course. Could the great artist have dreamed in his wildest fantasies, as he watched the comet in 1301, that someday a manmade device bearing his name would approach that very comet in order to study it at close hand?

NOTE: *By the time this book appears, Comet Halley will have circled the Sun, will be moving outward again, and will have had its encounter with the various probes. You can be sure I will have written another essay by then that will eventually appear in the third volume of this series.*

62 THE LARGEST SATELLITE

RECENT JUPITER PROBES have shown us how remarkable Jupiter's four large satellites are. The innermost, Io, is completely dry, has active volcanoes, and a surface of sulfur. The next, Europa, is mostly rock, but has a worldwide ocean and is covered by a smooth, unbroken shell of ice. The two outer satellites, Ganymede and Callisto, are a mixture of ice and rock and are covered with craters. The largest of the four large satellites is Ganymede, which has a diameter of 3,240 miles, one and a half times the diameter of our Moon. Can there be any other satellite in the Solar System to compare with these in size and interest?

The answer is yes. Indeed, the most remarkable one of all remains to be studied. It is Titan, the largest satellite of Saturn and the largest known satellite in the Solar System. Its diameter is 3,600 miles, which is 10 percent greater than that of Ganymede. Titan, in fact, is almost as large as the planet Mars. Nevertheless, for all its size, Titan doesn't have as much mass as Ganymede has. If one could weigh Titan and Ganymede, one would find that despite the larger size of the former, it weighs only 94 percent as much as Ganymede. This must mean that Titan is made of lighter and less dense materials than Ganymede. It may contain more ice and less rock than Ganymede does; it may also contain a considerable quantity of material that is still less dense than ice.

Back in 1944, an astronomer named Gerard Kuiper was able to show that Titan had an atmosphere. It is the only satellite in the Solar System that is known to have an atmosphere consisting of more than the tiniest wisps of gas. In fact, Titan's atmosphere is a substantial one that may be as dense as that of Mars and therefore 1/100 as dense as that of Earth. Titan's atmosphere is quite different from any other atmosphere we know. It seems to be made up almost entirely of methane, which has a molecule composed of one carbon atom and four hydrogen atoms (CH_4). On Earth, methane is the chief component of natural gas.

Considering Titan's size and assuming that its atmosphere is as dense as that of Mars, it turns out that there are about 53 trillion tons of natural gas on the satellite. That's about 1,600 times as much natural gas as there is thought to exist on Earth. What's more, if methane exists there, other related substances must exist as well. The carbon atom has the ability to attach itself to other carbon atoms to form chains and rings and does so at the slightest opportunity. For instance, careful analysis of the light reflected from Titan shows that there are small amounts of other gases present. These may be ethane or ethylene, and both these gases have molecules containing two carbon atoms.

If a two-carbon chain can be formed, why not chains that are longer still? Why not chains made up of three or four or seven or ten or more carbon atoms? This is quite possible—even inevitable. And what makes that interesting is that when we are talking of chains of seven or eight carbon atoms, with hydrogen atoms attached, we are talking of gasolines. Still longer chains make up kerosene and fuel oil and asphalt. Such "hydrocarbon" molecules make up substances that are only three-fourths as dense as ice, and this may help account for Titan's low density. In short, underneath that atmosphere of natural gas, Titan may have a gasoline ocean lapping up against a sludgy petroleum shore. If this is so, then Titan is one of a kind. There's nothing else like it in the Solar System.

But *is* this so? We've had one probe pass by Saturn and when the occasion was right, its sensors were pointed toward Titan. Unfortunately, at the moment when the probe was to send back signals of its findings, a mixup allowed some other satellite, which was lined up in the same direction, to send out interfering signals. It was a dreadful disappointment, but a second probe is on its way and with it, a second chance.

To be sure, it is not quite practical to look at Titan as an enormous oil well for Earth's use. Even when it is at its closest, Titan is 780 million miles away from us, or thirty-three hundred times as far away as the Moon. There's not much chance, therefore, of importing Titanian crude profitably. On the other hand, in centuries to come, when the outer Solar System has been colonized and settled, there may be occasions in special places when nuclear and solar energy are unavailable. In those cases, Titan may serve as a useful energy source, provided ample oxygen can be obtained in which to burn the gas and oil.

NOTE: *This essay appeared in December 1980. The second probe I referred to sent back its results at just about the time of that writing, and what it discovered outdated some of my statements at once. See the next essay, which appeared in April 1981.*

63 THE INVISIBLE GAS

IN THIS SERIES of essays, I routinely get out on a limb in my predictions and speculations of the future, knowing full well that as the future actually unfolds, some of my limbs may get sawed off. I never really expected, though, that any limb would be sawed off even as an essay got into print. That happened in the case of the previous essay. Just as it appeared, Voyager I, swooping past Saturn, turned in data concerning its largest satellite, Titan, and at once I could hear the sound of limb sawing.

I had said that Titan was the only satellite we knew of that had an atmosphere. This is still true! I also reported on what seemed to me to be the astronomical consensus that "Titan's atmosphere is a substantial one that may be as dense as that of Mars and therefore 1/100 as dense as that of Earth." Not true! Way off! The radio signals from Voyager I to Earth were, at one point in the probe's orbit, able to skim by Titan so closely as to pass through its atmosphere and emerge again. Some of its energy was absorbed, and from the details of that absorption, it could be calculated that the portion of the atmosphere that was traversed reached a density at its bottom that was 1.5 times that of Earth's atmosphere! That's one hundred and fifty times *more* dense than the atmosphere of Mars. And the radio signals may not have hit bottom, so the atmosphere may be still deeper and may grow even more dense.

With an atmosphere that unexpectedly deep, the solid globe of Titan is smaller than we thought. I had said that Titan was a little larger than Jupiter's largest satellite, Ganymede, and that, therefore, Titan was "the largest known satellite" in the Solar System. I even used "The Largest Satellite" for the title of my essay, for goodness sake. Well, if we subtract the atmosphere, the solid globe of Titan is a little smaller than that of Ganymede after all, and it is no longer the largest known satellite.

How is it possible that astronomers committed such an oversight? Well, the simpler gases that may possibly be present in an atmosphere in sizable quantities — such as hydrogen, helium, nitrogen, oxygen, and argon — do not absorb much in the way of visible light, infrared light, or even radio waves. Astronomers analyzing what is left over when radiation passes through a world's atmosphere are apt to miss the presence of such gases because the gases leave no mark, so to speak. Carbon dioxide, water vapor, ammonia, and methane, which have more complicated molecular structures and which can be present in atmospheres in significant amounts, *do* absorb certain fractions of the radiation and are easily detected.

For instance, early on carbon dioxide was detected in the atmospheres of Venus and Mars, but it was only when close-up studies were made with probes that the invisible gas, nitrogen, was found to be also present. The atmosphere of each world is about 95 percent carbon dioxide and 5 percent nitrogen. In the case of Titan, methane was detected thirty-five years ago, and it was that alone which seemed to have the density of Mars's atmosphere. The actual amount of methane seems to be at least three times that, but when preliminary efforts at judging the atmosphere of a small world some 800 million miles away are off by only a factor of three, that is pretty good.

It turned out, however, that Titan had not small quantities but rather massive quantities of nitrogen, which, at Titan's distance, remained invisible to earthly astronomers. From the close-up view of Voyager I, however, it was possible to deduce that Titan's atmosphere was composed of about 98 percent nitrogen and 2 percent methane (subject to further refinement in later, more precise evaluations). That means that the possible gasoline oceans and petroleum sludge I speculated about in the earlier essay may still exist. Nitrogen liquefies at even lower temperatures than methane does, so we can have methane lakes with gasoline dissolved in them under nitrogen gas. In the upper, colder regions of the atmosphere, even nitrogen may liquefy, but the nitrogen rain may evaporate as it enters the warmer, lower reaches of the atmosphere and may never reach the ground.

What ground? The atmosphere of Titan is filled with tiny particles of as yet unknown composition. The particles may be tiny drops of liquid nitrogen or bits of chemicals similar to those found in automobile exhaust smog. Whatever they are, they cloud the atmosphere and do not permit us to view the ground.

Perhaps we could perform a radar scan, as was done in the case of Venus, or send probes down through the atmosphere to the surface, as was done in the case of Mars. Voyager II, however, which is on the way, isn't equipped to do anything that will reveal the surface, and after Voyager II, there is nothing planned for Saturn. Titan remains by far the most interesting satellite in the Solar System, but planetary exploration is winding down toward a halt, and our curiosity, having now been aroused, must go unsatisfied.

IN THE EARLY 1800s, Uranus, the seventh planet from the Sun, was the farthest one known. Its motion in its orbit around the Sun was not quite right, however, if Newton's law of gravitation was correct (and all the astronomers were sure it was). One reasonable explanation for the inconsistency was that there was a large planet lying beyond Uranus that had not yet been discovered and whose gravitational pull had not been allowed for; and it was that gravitational pull that was forcing Uranus to move not quite according to calculations.

Astronomers used Newton's theory to calculate where a planet ought to be to account for the discrepancy in Uranus's motion. In 1846, one of them looked in the neighborhood of the calculated position, and what do you know? The unknown planet was indeed there! That was how the planet Neptune, eighth from the Sun, was discovered. It was the greatest victory for Newton's theory in all its history. Once Neptune had been located, its gravitational pull was allowed for and Uranus was seen to follow the calculations much more closely—but not with complete exactness, either. There still remained a tiny discrepancy, a hardly noticeable one.

Could there be a planet beyond Neptune that accounted for that last little bit of error? Such a planet would be so distant that its gravitational pull would be very weak and would introduce that tiny error. It would be nice, astronomers thought, to find that missing planet and settle everything. The trouble is that the farther a planet is from the Sun, the dimmer it's bound to be, and the more difficult it is to find it amid the crowds and crowds of equally dim stars surrounding it on all sides. To be sure, a planet can be recognized as such by the fact that it slowly moves, relative to the surrounding stars, but the farther away it is, the more slowly it appears to move and the harder it is to spot that motion.

For decades, the planet beyond Neptune (number nine) was searched for without any success. It wasn't till 1930 that it was finally detected — very dim and moving very slowly — close to the spot where calculations showed it ought to be. That created another sensation, but planet number nine, which was named Pluto, turned out to be far dimmer than expected. Even allowing for its distance, astronomers realized that it couldn't possibly be as massive as Neptune. They thought it might be no more massive than Earth (which is only about one-fourteenth as massive as Neptune). Over the years, further investigations showed that it couldn't be even as massive as Earth and that it might be no more massive than Mars (which is only about one-tenth as massive as the Earth).

Then, in 1978, it was discovered that Pluto had a satellite, which was named Charon. By studying the distance of Charon from Pluto, along with the time it took for Charon to revolve about Pluto, the total mass of Pluto and Charon could be calculated at last. It turned out to be only about one sixty-fourth the mass of Mars. Pluto is actually no more than a large asteroid.

With Pluto-Charon that small, astronomers concluded, its gravitational effect should be small, too — far too small to have any noticeable effect on Uranus. Could it be that Pluto-Charon was not what astronomers were looking for in the first place? That its location in the right spot was just coincidence? If so, there must be a planet number ten, farther off still, but much more massive than Pluto, and *that* would be the object we had been searching for all along. Yet astronomers hate to go back to their calculators without something better to work on. The discrepancy in Uranus's motion is too small to give them good figures. It would help if we could study Neptune's motion in detail, since if the suspected tenth planet is out there, its effect on Neptune (which is nearer to it than Uranus is) should be greater than on Uranus. To find the exact error in Neptune's motion, however, it would help if it were studied for several revolutions about the Sun. Unfortunately, it takes that planet 165 years to make one revolution, so in all the time since it was first discovered, it has not yet had time to go completely around the Sun even once.

In 1982, though, it was discovered that nearly four hundred years ago, the Italian scientist Galileo observed what he thought was a star that was almost exactly where Neptune was at that time. It probably *was* a planet. Galileo did not find it precisely where theory predicts it should have been, though. His position was off by one-thirtieth the width of the Moon. That isn't much, but it's a lot to astronomers. If Galileo's observation holds up, it may be possible to estimate the error in Neptune's movements quite well and then make a good calculation as to where planet number ten ought to be. In that case, the search could begin in earnest, with the full armory of modern astronomical instruments, including the use of probes sent into the outer Solar System. How interesting it would be to spot planet number ten, a giant world circling the Sun far out in the cold; a distant planetary cousin whose dim, mysterious gloom we could finally reach out and touch.

65 THE DOUBLE STAR

THE STAR NEAREST to us is Alpha Centauri. It is only 4.3 light-years away. That's not exactly next door, for that distance is equal to 25 trillion miles, which makes it a quarter of a million times as far away from us as our own Sun is. Just the same, all other stars are farther away from us than Alpha Centauri is. What's more, as best we can tell from this distance, Alpha Centauri is an almost identical twin to our Sun. It is just about as large and as hot as our Sun and has very nearly the same chemical constitution. The next nearest star that so closely resembles the Sun is about six times as far away as Alpha Centauri.

It would seem, then, that if ever we try to explore some other star, the one we should surely aim for is Alpha Centauri. Not only is it the closest star, but if it is so like our Sun, it may have planets very much like those of our Solar System, one perhaps that is rather Earth-like and that bears life. There is a catch, however, and it may be an important one. Alpha Centauri is not a single star, as our Sun is. It is a "binary star," a system of two stars that swing about each other—a double star, so to speak. The Alpha Centauri I've been speaking of as our Sun's twin is, actually, Alpha Centauri A, the brighter of the pair. The other one, Alpha Centauri B, is smaller and cooler and is only about a quarter as bright as Alpha Centauri A (or our Sun) is.

Actually, being a binary isn't terribly unusual among stars. When such binaries were first discovered a little over a century and a half ago, it was thought that there were few of them. The more closely the stars were studied, however, the more of them turned out to be binaries. Right now, it would appear that well over half the stars we see are systems of two (or sometimes more) companions. It is the stars who live in single blessedness, like our Sun, that seem to be the more unusual ones.

This brings up an important point. Some astronomers feel that when stars form in pairs, they aren't likely to form planets also, because planets would not have stable orbits with two stars pulling at them. Only single stars, according to this view, would be likely to have planetary systems. This would cut down the chances for life in the universe by a good deal, and there would be no hope for life in the Alpha Centauri system.

But is this view a fact? Here in our Solar System, we have planets like Jupiter and Saturn, which are large enough to be considered almost starlike and which both have a number of satellites circling around them and making up miniature "solar systems" of their own. Might it not be, then, that Alpha Centauri A and Alpha Centauri B each has a planetary system of its own?

If Alpha Centauri B were part of our own Solar System and circled our Sun instead of circling Alpha Centauri A, its elliptical orbit would carry it, at one end, nearly as close to our Sun as Jupiter is. At the other end of its orbit, it would be farther away than Neptune is. The outermost planets, from Jupiter onward, would not be able to move in their present orbits under such conditions, but they are too far from the Sun to bear life anyway. The inner planets of the Sun, however, including Earth, would not be affected much by Alpha Centauri B.

Alpha Centauri B would never be closer to Earth than about half a billion miles, so its gravitational pull would not bother us. Nor would it be close enough to look like a second Sun, for it wouldn't show as a visible circle of light. As it circled our Sun in about eighty years, it would always have a starlike appearance. It would be a very bright star, of course. Even at its farthest distance from Earth, it would be one hundred times as bright as the full Moon, and forty years later, when it was in its closest proximity to Earth, it would be fourteen hundred times brighter than the full Moon. Even at its brightest, though, it would be only $1/325$ times as bright as the Sun. It would deliver a little light and heat, varying in quantity from decade to decade, but not much. Earth should get along just as well with Alpha Centauri B as without it.

It may be possible, in fact, for there to be *two* Earth-like planets in the Alpha Centauri system, one circling each of the two stars. And maybe both planets bear life of some sort. You see, then, that Alpha Centauri, instead of presenting us with no life because it is a binary system, may present us with two samples for the price of one. And imagine the possibilities if two worlds within less than a billion miles of each other both had intelligent life. I think we are very fortunate that the star which happens to be nearest to us also offers us such exciting possibilities.

66 PLANETS IN BIRTH

IN MY ESSAY "Orbiting Telescopes" (see *Change!*, Houghton Mifflin Co., 1981), I said that a large telescope in space, doing its viewing from beyond Earth's interfering and troublesome atmosphere, would very likely make important new discoveries. Well, it's happened. An Infra-Red Astronomical Satellite (IRAS) has been launched, one that is equipped to detect and report on infrared radiation from space (radiation that is not quite energetic enough to be visible to the eye). Such radiation is largely absorbed by our atmosphere and cannot easily be studied from Earth's surface.

IRAS detected infrared radiation from the star Vega. In itself, this is not surprising, for just about all stars (including our own Sun) radiate copiously in the infrared. Vega, twenty-six light-years away, is the fifth brightest star in appearance. In actual fact, it is twice as large as the Sun and shines sixty times more brightly, so why shouldn't it send out lots of infrared? The catch is that even allowing for Vega's size and brightness, the infrared radiation was far more intense than expected. Astronomers took a closer look and, behold, the infrared radiation was not coming from Vega itself but from the regions of space nearby. Vega, it seems, is surrounded by a shell of something-or-other that radiates in the infrared, a shell that is about 7.5 billion miles thick. The shell stretches twice as far from Vega in every direction as Pluto's orbit about our Sun, so it is a shell that is considerably larger than our planetary system.

The shell isn't *very* hot; it doesn't take much heat for material to emit infrared radiation. It is about as hot as the rings of Saturn. Judging from the nature of the radiation, astronomers suspect that the shell consists of particles that are too large to be classified as dust, particles possibly pinhead-size or larger. In fact, it's suspected that Vega is in the process of forming a planetary system of its own. It is still in the early stages of the process, as it is less than 1 billion years old (as compared with the Sun's age of nearly 5 billion years). Furthermore, it makes sense to suppose that Vega is a bit slower about forming planets than our Sun was, for the following reason.

All stars send out a continuous stream of subatomic particles in all directions. This is called the "stellar wind." The larger and brighter a star is, the stronger its stellar wind. This wind keeps the small particles surrounding a star stirred up and tends to delay their coalescence into planets. Vega, large and bright as it is, would have a far stronger stellar wind than the Sun would have had at a similar stage in its history. The shell of particles about Vega would therefore be less well coalesced than a similar shell about our Sun would have been at a similar time in its history. Even so, we needn't suppose that Vega's shell of particles contains only small bits of gravel and nothing more. It might contain quite a few larger bits and even a handful of planetary-size objects that are in the slow process of gathering up all the smaller objects as they circle Vega. It so happens that we would detect the smaller objects but not the larger ones.

That may seem paradoxical, but it isn't. During a light rain, we have no trouble seeing clearly; the raindrops don't scatter much light. If the same amount of water is divided up into much tinier drops — drops so small they simply hover in the air — then we have millions of times as many droplets that can scatter far more light than the few larger drops would. We then find ourselves surrounded by the impenetrable whiteness of a fog. Similarly, an amount of matter divided into trillions of tiny bits would radiate far more infrared than that same amount of matter collected into large clumps.

It seems reasonable, then, to suppose that Vega has a planetary system already in existence, one that is still in the process of growing. It is the first clear case of a star other than our Sun for which this is true. (Over the past few decades, some nearby small stars have been reported to show wobbles in their motion that could be the result of the gravitational influence of large planets, but the data were very borderline, and many astronomers disputed the hypothesis. The case of Vega seems much firmer.)

If Vega, as well as our Sun, has a planetary system, then it is now much more plausible to suggest that other stars also have planetary systems than it was when our own solar system was the only known instance. For this reason, the discovery has given heart to those astronomers who have been reasoning all along that the formation of planetary systems is a natural stage in the evolution of stars. If this is true, it helps provide one of the logical steps that is necessary to the argument put forth by some astronomers that life, and even technological civilizations, may be quite common in the universe. We see, then, that an unexpected discovery by a telescope in space can encourage us in our search for extraterrestrial intelligence.

67 IN BETWEEN

IF ONE IS SPECULATING as to whether there is life elsewhere in this Galaxy, it is important to know whether planets are a common phenomenon. Our Sun has a system of planets, of which Earth is one, but do other stars have such systems? If there are no planets circling other stars, it seems unlikely that there is any life elsewhere. Astronomers are almost certain that other planetary systems exist and that they are even common, but it is almost impossible actually to *detect* planets outside our own Solar System. To be sure, a belt of dust has been recently located about some stars (see the previous essay), which may indicate planets in the process of formation, but that is not quite the same thing.

The elusiveness of planets is not surprising. Planets are much smaller than stars, so small that the temperatures and pressures at their centers aren't large enough to ignite the fusion reactions that give stars their light and heat. This means that planets can be seen only by the reflected light of a nearby star, and this very dim light is made imperceptible by the blaze of the nearby star itself. However, other techniques are possible. Held together by gravitational pull, a planet and its star circle each other. The planet makes a larger circle, and the much more massive star makes a much smaller circle. In the case of the Sun and Earth, the difference in mass is such that the Sun's motion, as Earth circles it, is unnoticeable. However, if the star is unusually small and the planet unusually large, the star may make a noticeable circle. Ordinarily, such a star moves very slowly across the sky in a straight line, but the pull of a large planet would make it "wobble" along its course.

Such wobbles would be observable if the star were near enough to us.

In the past forty years, astronomers have reported such wobbles among about half a dozen small and nearby stars. From the size of the wobbles, it is possible to estimate the size of the planets and, occasionally, even to calculate the presence of two planets. The trouble is that the wobbles are barely perceptible and the figures on the supposed planets are highly dubious. As the years have gone by, more and more astronomers seem to have decided that the data derived from the supposed wobbles are not reliable.

But now there is the case of a small star named Van Biesbroek 8 (named for the first astronomer who studied it carefully), more easily referred to as VB8. It is about twenty-one light-years away from us, in the constellation of Ophiuchus. In 1983, three American astronomers, after meticulous measurements, have reported a wobble in the star's slow progress across the sky. What makes this wobble different from earlier ones? Well, astronomers have developed new techniques. One of them, "speckle interferometry," uses infrared light, which makes it possible to detect small objects very close to stars. In 1984, a tiny object was detected near VB8, and it is this object that may account for the wobble. If so, this would be the first planet actually detected outside our own Solar System. It's not the kind of planet we're used to, however. The planets of our Solar System are all cold and dark on the outside, except where they are warmed and lit by heat and light from the Sun. BV8's planet, however, seems to have a temperature of nearly two-thousand degrees Fahrenheit, so its surface shines with a red hot glow.

Why should it be red hot? Well, when planets form, small fragments collide and cling. More and more small fragments strike one another, and the whole compresses under its own gravity. The motion of the fragments is converted to heat once they collide, and the compression increases the temperature further. Thus, Earth is still very hot at its center. The larger the planet, the hotter the center grows, so that Jupiter is much hotter inside than Earth is. An object much more massive than Jupiter develops so much heat at the center that nuclear fusion begins to take place (like setting off a hydrogen bomb) and the object becomes a star. VB8's planet is considerably more massive than Jupiter, apparently, but not massive enough to form a star. It is, however, massive enough to have a gravitational pull sufficiently large to compress it into a somewhat smaller size than Jupiter. While this does not develop enough heat to start nuclear fusion, it *does* develop enough heat to keep the surface red hot.

What do we call it, then? It seems odd to call it a planet if it differs from all the planets we know in such a spectacular way. After all, it *shines,* and planets don't shine. On the other hand, it would be peculiar to call it a star, since the shining isn't the result of nuclear fusion. It is something in between. Some astronomers refer to it as a "brown dwarf" — "brown," because it isn't quite dark enough to be black, and "dwarf" because it is quite small for a star. I prefer to think of it as a "substar." But whatever it is, it seems to be something we've never encountered before.

68 THE NEXT EXPLOSION

EVERY ONCE IN A WHILE, a star appears in the sky that wasn't there before. Usually nobody notices, because few people stare at the sky these days and because those who do haven't memorized the pattern of the stars. Astronomers notice them, of course, especially amateur astronomers who do nothing but sweep the sky with their instruments in search of anything that might be interesting.

In ancient times, such new stars were considered exactly that: new stars. Temporary ones, too, for they always faded away after a few weeks or months. In Latin, "new star" is "nova stella," and such a star has been referred to as a "nova" ever since. Such stars, however, are neither new nor temporary, as was discovered once the telescope was invented. The star was there before it became a nova; it was just too dim to be visible to the unaided eye. Then suddenly, within a day or two, it would brighten by several tens of thousands of times and be bright enough to see by eye. Slowly, though, it would fade, until it was once again too dim to see without a telescope.

In some cases, though, a star brightened not by tens of thousands of times but by hundreds of millions of times. Such a star ends up not just another star in appearance. At its brightest, it can be bright enough to outshine the planet Venus, bright enough to cast a shadow; bright enough to be seen by daylight. Such a star was visible in 1572 in the constellation Cassiopeia. For weeks it outshone everything in the sky but the Sun and the Moon. Only thirty-two years later, in 1604, another very bright nova appeared. This nova wasn't quite as bright as the one in 1572 — it was only as bright as the planet Jupiter.

Earlier than that, in 1054, a star had appeared that was every bit as bright as the later star of 1572, but astronomy was at a low ebb in the West in that century and, except for a couple of very obscure and uncertain references, there is no European report on it. The star lit up the sky like a super-Venus in the constellation Taurus, but Europe seems to have been oblivious of it. How do we know about it, then? Ah! Astronomers in China kept meticulous track of such things and we have *their* reports.

Astronomers have found objects in the sky which seem to show that in about 11,000 B.C., before what we call civilization had gotten started, an even brighter star appeared in the far southern constellation Vela. At its brightest, this star may have appeared as bright as the Moon. Such particularly bright stars are not just novae; they are "supernovae." A nova represents a relatively minor explosion of matter on the surface of a "white dwarf" star, a tiny star with enormous gravity. A supernova, on the other hand, is a star that explodes altogether, driving most of its matter into space. In the process, it blazes up to a brightness equal to that of a billion or more normal stars. Astronomers would love to be able to study a supernova in detail, but since 1604—for nearly four hundred years now—there has not been a single supernova in our own neighborhood of the Galaxy. The enormous technological batteries of astronomy have had nothing to work on. There have been supernovae, yes, but only in distant galaxies, thousands and even millions of times as far away as the great nearby supernovae of 1572 and 1604. But we know a good deal now about how stars evolve and what precedes the giant explosions. We ought to be able to study the stars of our neighborhood to determine whether any is approaching the explosive stage, and get set to follow that next explosion.

One particular star has attracted attention. It is called Eta Carinae and is an enormous star, a hundred times as massive as the Sun. It is a red giant, which is the last stage before explosion. It is surrounded by a cloud of debris, which is an indication of gathering instability. And for a long time it has been pulsating, undergoing minor explosions that have changed its brightness now and then. In 1840, it brightened to the point where it became the second brightest star in the sky. Then it dimmed, and nowadays it is too dim to see with the naked eye. It delivers an extraordinary amount of infrared radiation, however, that is not visible to the eye. Finally, astronomers have recently detected nitrogen in the cloud surrounding Eta Carinae. Nitrogen would ordinarily be located well below the surface of the star, so obviously Eta Carinae's substance is really being stirred up. Astronomers now think that Eta Carinae will be the next star to explode; in fact, they consider it to be on the point of going.

There are two catches. First, when an astronomer says "on the point of going," he or she means any time within the next ten thousand years or so. And second, Eta Carinae is so far down in the southern sky that when it does go, the explosion won't be visible from Europe or from most of the United States.

NOTE: *This essay appeared in January 1983. Since then, astronomers have been studying Eta Carinae, and it seems to be even more massive than had been suspected. It is probably two hundred times the mass of the Sun, and it belongs to a new class of "very massive stars" which, until fifteen years ago, astronomers had assumed couldn't possibly exist. The peculiar behavior of Eta Carinae may be normal for such superstars and may not indicate the imminent coming of a supernova (though it also may, of course).*

69 WHERE THE PEOPLE ARE

SUPPOSE THAT SOME INTERPLANETARY explorers approached Earth some thousands of years ago and viewed it from a great distance and tried to figure out where the people might be so that they wouldn't land in a wilderness. It would have been a good bet for them to explore the course of some prominent rivers. A river is a good communication artery for people with a primitive technology and offers a way of irrigating the soil to ensure an abundance of food. And indeed the very earliest civilizations sprang up along rivers such as the Nile, the Euphrates, the Indus, and the Hwang Ho.

Suppose the explorers had come along in the last century and had been interested in finding the most advanced regions in order to waste as little time as possible in establishing trading posts. They might have decided to search for sites where the resources necessary for industrial and technological development were present in cheap abundance. They would have looked for coal and iron deposits. And, indeed, the Industrial Revolution was initiated and underwent its early expansion in those regions where coal and iron were easily obtained.

Now astronomers on Earth are planning to make long-distance explorations. They hope to search the sky for signs of the existence of possible distant civilizations. Where ought they to look? The obvious answer is that they should look in the neighborhood of stars that are rather like the Sun — single stars that are of medium mass and medium age. Earth-like planets may be circling those stars and may prove to be the home of life, intelligence, and civilization.

But they may be only ordinary civilizations. Suppose we wanted to find the most advanced civilizations possibly in existence and decided to look, therefore, in places where a great deal of energy might easily be available. A great deal of energy could be obtained in the neighborhood of black holes, especially of those that aren't so large as to be impossibly dangerous — say, no more than star-size black holes. The trouble is we're not certain where such things are, or even *if* they are.

How about pulsars, though? These were discovered in 1969, but already we have found hundreds of them and, in all likelihood, many thousands of them exist scattered over our Galaxy. Pulsars are remnants of huge supernova explosions. They are collapsed stars. All the mass of an ordinary star like our Sun is compressed into a tiny ball perhaps no more than eight to ten miles across. Such a star has an unimaginably intense gravitational field. A pulsar rotates about its axis in a few seconds, sometimes in a few tenths of a second; and, in at least one case, even in a few hundredths of a second. Pulsars have immense rotational energy. The magnetic field of a pulsar also collapses as the star does and is compressed to enormous intensities.

Now imagine a huge conducting device placed in orbit about a pulsar. It would continually cut through the intense, magnetic lines of force, and a vast electric voltage would thus be set up in it. It would be an inconceivably large electric generator, a power station far larger than anything we could build about our Sun. Suppose many such devices were set up in orbit at varying distances from a pulsar. Unlimited quantities of energy might then be obtained from the tiny star. The acquisition would be made at the expense of the rotational energy of the star, and the star's rotation would gradually slow, but it would be billions of years before the slowing would become appreciable.

Of course, pulsars are hundreds of light-years away, at the very least. We're not likely to reach any of them for many centuries. Suppose, though, that other, older civilizations have already managed to reach various pulsars and have set up devices for tapping their enormous energies. These civilizations could then construct large numbers of space settlements about the pulsar that would make use of all that energy. Those space settlements may already exist, and they may represent the most important concentrations of advanced technology anywhere in the Galaxy. They may be "where the people are," and, if we want to search for some sign of civilization, it might be sensible to study the neighborhood of the pulsars.

There is, unfortunately, a catch. (Isn't there always?) As the pulsars turn, in seconds or fractions of a second, they spew out high-speed particles and all sorts of radiation from their magnetic poles. These streams sometimes pass in our direction as the pulsars turn, so that we receive periodic pulses of particles and radiation. These are signals, too; such intense signals that they would probably obscure any signals that arose from a mere human technology. Of course, we might choose those pulsars whose pulses miss us as the little stars turn, but then we couldn't identify them as pulsars. Too bad.

NOTE: *Credit must be given where credit is due. The idea on which this essay is based is not the product of my own thinking; it was suggested by a friend of mine named Mark Berry.*

70 INVISIBLE ASTEROIDS

IN AN EARLIER ESSAY of mine, "The Cosmic Subway Line" (see *Change!*, Houghton Mifflin Co., 1981), I wrote about black holes. These are conglomerations of matter packed together so densely that the gravitational fields in their immediate vicinities are enormous, so much so that nothing that falls in can ever get out again (hence "holes"). Even light can't emerge (hence "black"). In the present-day universe, the only cosmic event forceful enough to form a black hole would be the sudden collapse of a massive star near the end of its life cycle, when it ran out of nuclear fuel to keep itself expanded. Such a star would have to be at least three times as massive as our Sun.

Consequently, all the black holes that have formed over the whole period of billions of years in which stars have existed have masses that are star-size. Some of the black holes have managed to grow by swallowing matter from other stars. If the black holes are massive enough, they even become capable of swallowing entire stars at a gulp. It is quite possible, therefore, for a black hole to have a mass equal to that of a million stars. Galaxies are made up of as many as hundreds of billions of stars, and some astronomers suspect that there are huge black holes with masses of a million stars at the center of every galaxy, including our own. Globular clusters, made up of a few hundred thousand stars, may have black holes with masses the equivalent of a thousand stars at their center. There may even be ordinary star-size black holes scattered here and there in space.

The trouble is that all these black holes are hard to detect and identify. Even when they are not well hidden by crowds and crowds of normal stars, they are black and therefore not visible against the black sky. Black holes reveal their existence only through the x rays that are emitted when matter falls into them —and, of course, those x rays may exist for other reasons. The very existence, let alone the properties, of black holes therefore remains controversial, though most astronomers believe they *do* exist and that some *have* been detected. There is a kind of black hole, however, that is particularly elusive.

The British physicist Stephen Hawking pointed out some years ago that there is another way in which black holes might have been formed. About 15 billion years ago, the universe came into being. All its mass was apparently concentrated into one location at that long-ago time, and it exploded. From that explosion, or "big bang," there gradually developed the universe as we know it. In the enormous temperatures and pressures of those first moments after the big bang (far beyond anything that now exists), bits of matter may have been squeezed into black holes, Hawking argues. Furthermore, the black holes then formed did not have to be star-size, as those formed nowadays must be. They could be of all sizes, and some quite small ones may have been formed—no more massive than planets or even asteroids, or smaller still. These are "mini–black holes," and if star-size black holes are hard to detect, asteroid-size black holes should be even more so, one might suppose. An asteroid shrunken to the size of a pinhead or smaller would certainly be invisible, even if it were located only a few million miles away within our Solar System.

Well, perhaps not so. Hawking applied the rules of quantum mechanics to black holes and decided that they *do* give up matter; not in large chunks, but as subatomic particles. By giving up such particles, black holes slowly "evaporate." The more massive a black hole, the more slowly it evaporates, and a star-size black hole is bound to pick up matter at a rate more rapid than that of its excessively slow evaporation as it moves through space. That kind of black hole would then continue to grow inexorably, under present-day conditions.

A mini–black hole, however, picks up less matter because, being less massive, it has a smaller gravitational field. It also evaporates more quickly. Evaporation could easily outweigh growth, and many mini–black holes would tend to shrink in mass as time passed. The smaller they became, the faster they would shrink, until—when they got really small—they would go all at once and produce a sudden burst of very energetic gamma rays. Some mini–black holes that were produced at the time of the big bang were so small to begin with that they have long since evaporated. Some that were considerably larger still exist today and will continue to exist for a long time.

Mini–black holes that happened to be just the right size at the time of the big bang, are undergoing their final evaporation now. Astronomers believe they know exactly the characteristics of the gamma rays that should be produced, and they are watching for that "signature." They haven't been watching very intently or for very long, and so far, they haven't detected any gamma ray signatures. It may be that mini–black holes don't exist. If they do, though, they may offer a hazard to space flight.

71 TICKING 642 TIMES A SECOND

FOR SCIENCE TO PROGRESS, it is absolutely necessary to be able to measure time—the more accurately the better. The first clocks men had were astronomical. Earth's turn about its axis measured the day; the Moon's phases measured the month; the

Sun's rise and fall at noon measured the year. None of these things, however, could measure the parts of a day. Human beings tried to follow the shadow of the Sun or to note the burning of candles or the movement of sand or water through a small hole. That might get you to the nearest hour. Galileo, in his early experiments concerning motion, had to use his pulse as a clock, or the dripping of water. It's a wonder he got the right answers.

It was not until 1656 that the pendulum clock was devised and that, for the first time, people had a way of measuring time to the nearest minute. It was an age of long voyages, and there was no way of measuring longitude without an accurate clock, but pendulum clocks wouldn't work on the swaying deck of a ship. Not until 1765 were the necessary chronometers built, and then, for the first time, ships could work out exactly where they were on the featureless deep.

Clocks have continued to improve steadily since then, and we now have "atomic clocks," devices that measure the vibration of atoms in their changeless rhythm. By counting those vibrations, one can measure time so accurately that it is easy to see that Earth's rotation on its axis (the original clock) is not at all steady. Because of earthquakes and the consequent shifting of rocks under our feet, together with the movements of the atmosphere with the seasons, the Earth's rotation speeds and slows at odd times. On June 30, 1985, it was necessary to add a full second to the day in order to keep Earth in line with the atomic clocks.

An atomic clock is an excellent device, but is it conceivable that we could turn back to the heavenly bodies to tell time exactly? Could it be that there is some clock in the sky that is more accurate—*much* more accurate—than anything on our raggedly turning Earth? Every heavenly object turns about its axis and moves through space, but the turning usually involves hours, and the other motions involve anywhere from months to many millions of years. What we need is something that completes some sort of cycle in a second or less and does so with enormous regularity. Such an object would itself have to be very small, and yet it would have to signal its cycle across the light-years.

Hopes rose when pulsars were discovered. They are tiny "neutron stars" no more than eight to ten miles in diameter, and they turn on their axis within anywhere up to four *seconds*, doing so with great regularity. To be sure, they tend to lose energy as they turn, and as their energy drains away, their rate of turning slows. The slowdown, however, is itself slow and regular and can be allowed for. The real shortcoming of the pulsar is that every once in a while there is a tiny but sharp and unpredictable change in its rate of turning. This is called a "glitch." We don't know what happens. It may possibly be a shift in its very dense material, a sort of "starquake." This would raise havoc with the use of pulsars as clocks.

But late in 1982, a pulsar was discovered that has an unusually rapid rate of turning. It completes its rotation in 1/642 of a second. To put it another way, it turns 641 times a second. As it turns, astronomers note the rise and fall of its microwave radiation, like a clock making 642 ticks a second. What's more, this "ticking" is extraordinarily regular. For some reason not yet quite understood, the rapid pulsar loses energy much more slowly than ordinary pulsars do, and during the time in which it has been observed, it has never once undergone a glitch. The ticking repeats itself with an accuracy that can be counted to sixteen decimal places. If the rapid pulsar were to keep going for 5 trillion years (three hundred times the present age of the universe), it might lose or gain a second.

What makes things even more exciting, is that a second rapid pulsar has been discovered, and it may be that we will end up with five or six of them. If they all behave as the first one does, and if astronomers can time them all, synchronize them, use each one as a check on the others, and combine them with atomic clocks here on Earth, the measuring of time might be made so nearly perfect as to require no improvement for any imaginable purpose. With such a nearly perfect clock, astronomers might detect tiny additional irregularities in the movements of Earth, of the Moon and of human-made satellites and probes, and of the planets and stars, too. The discovery of these irregularities of motion might give rise to a deeper knowledge of the structure of stars and planets, of the influences to which they are subject, and of gravitational theory and the universe itself. Like Galileo four centuries ago, we need only a better clock to understand so much more, and we are on the brink of having it.

---

72 MYSTERY OF
THE MISSING MASS

NEARLY 99.9 PERCENT of all the material (or "mass") making up our Solar System is packed into one body—the Sun. Everything else put together—the Earth, the Moon, the planets, the satellites, the asteroids, the comets, the meteors—makes up only a little over 1/1000 of the whole.

Astronomers felt they had every right to suppose that this was true of all other stars. Every star, they felt, must have far, far more mass than all the objects that circled it. Therefore, the stars scattered through the universe must, in themselves, contain virtually all the mass of the universe.

Stars, by their very nature, are hot and glowing and can be seen. Using modern instruments, individual stars can be seen at distances of many thousands of light-years, whole galaxies of stars at many millions of light-years, particularly bright galaxies, called "quasars," at many billions of light-years. Astronomers calculated the amount of this visible mass of the universe and decided how much there was in every cubic light-year; that is, what the average density of the universe was. From this density, they could calculate the overall intensity of the universe's gravitational field—the force that kept pulling the galaxies together. As it happens, the universe is expanding; the galaxies are flying apart as the result of the original big bang some 15 billion years ago. The universe's gravitational field, it turns out, is far too weak to stop the expansion and, someday, billions of years hence, to force the galaxies to begin coming together again into an eventual "big crunch." The universe, astronomers decided, would expand forever.

Then came problems. Galaxies rotated, and to the eye it seemed that almost all the mass in galaxies was in the center, where stars were very thickly spaced. If that were so, galactic rotation should be slower and slower as one moved out from the center. In recent years, astronomers succeeded in measuring the rate of rotation of faraway galaxies at different distances from the center. They found there wasn't the right kind of fall-off; the outskirts rotated too quickly. This seemed to mean that the mass wasn't concentrated in the center but was far more spread out into the outskirts of the galaxy and beyond. That additional mass, however, *could not be seen.*

Then, too, galaxies exist in clusters, with individual galaxies moving about within them like bees in a swarm. It's the gravitational fields of the individual galaxies of the cluster that keep the whole structure together. However, judging from the quantity of visible stars in the galaxies, there isn't enough gravitational intensity to do the job—yet the cluster holds together. There must be an additional mass in the cluster that we can't see. Astronomers began to speak of "the mystery of the missing mass."

Of what could the missing mass consist? It couldn't consist

of stars, since it is invisible. Perhaps it consists of objects like planets that are too small to shine. Could every galaxy have a halo of such small bodies stretching far beyond its visible borders and making each one a hundred times as massive as it seems to be? This didn't seem possible. The number of small bodies required for the purpose seemed far too great to be likely. Perhaps the answer was black holes. These could be even smaller than planets but could each contain as much mass as a star, even many stars. Perhaps every galaxy was surrounded by a halo of a not unreasonable number of black holes.

Yet there are theoretical reasons for supposing that any missing mass can't be made up of anything that is in turn made up of protons and neutrons. Stars, planets, and black holes are all made up of protons and neutrons, so the answer has to be something else. But what else is there?

It could be neutrinos. These are a billion times as numerous as protons or neutrons but are thought to have no mass at all. What if they have a very tiny bit of mass? Or what if there are particles that some theoreticians have proposed but which have never been spotted—magnetic monopoles or gravitinos or axions? It's got to be *something*.

Does it matter? Oh, yes, it does. Right now, it's beginning to appear that we know only about 1 percent of the universe. If we ever find out what the rest is made of, we are sure to have to revise our notions of how the universe began and of how it will end. The knowledge will confirm or destroy various notions that physicists have about the nature of matter, and it may enormously extend, or collapse, many basic beliefs in science. And who knows how such new knowledge may be applied to human life and so produce as yet unimaginable changes in society? We can only wait and see.